The Iron Major Survival Guide

David W. Dunphy

Graybeard Publishing

First Published in 2016 by Graybeard Publishing

Copyright © 2016 Graybeard Publishing

All rights reserved. No part of this publication may be reproduced or distributed in any form or by any means, electronic or mechanical, or stored in a database or retrieval system, without prior written permission from the publisher.

This is a work of nonfiction and all opinions expressed are solely those of the author. They do not reflect the official policy or position of the Army, Department of Defense, or the US Government. Any and all views expressed by any person in this work does not constitute an endorsement by the US Army, the Department of Defense or the US Government.

This book may mitigate the threat of douchebaggieitis.

ISBN: 978-0-9912382-8-6

Graybeard® is a registered trademark of Graybeard Publishing

Printed in the United States

Cover Design by Thomas Hunt

Edited by Kelly Crigger

For my wonderful kids: Kristina, Alek, and Dylan; and for my poor parents, COL(R) Donald and Kathleen Dunphy, who have had to bear the burden of me for far too long.

Acknowledgments

Much of this book was made possible through the contributions, guidance, and mentorship of a great many officers, non-commissioned officers, Soldiers, and civilians who, directly or indirectly, shaped my experiences, career perceptions, and leadership philosophies—for better or for worse. I would especially like to thank BG John Richardson, COL Matthew Zimmerman, COL(R) Mark McKnight, LTC(R) Paul Anderson, LTC(R) Steve Herring, and my old friend and publisher, LTC(R) Kelly Crigger for their azimuth checks and valuable insights into the development of this and earlier versions of The Iron Major Survival Guide. I also want to thank the good people at the Sierra Nevada Brewing Company for their hoppy concoctions that so nurtured my enthusiasm and philosophical debauchery through the composition of this fine work. Finally and most importantly, a heartfelt thanks to all great American warriors, past and present, who chose service to the Nation, the most noble and selfless of professions.

Preface

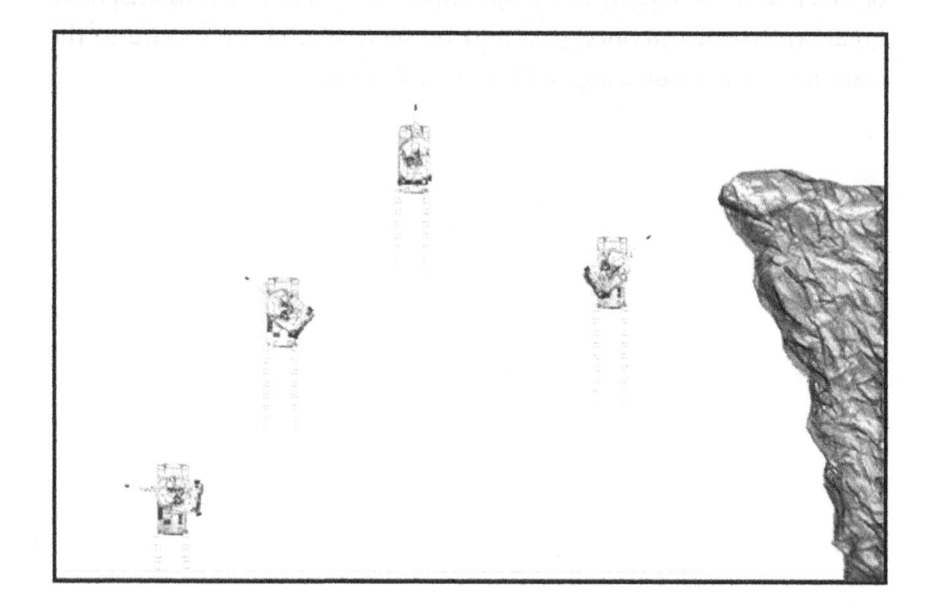

The black mass of the Whale rose up on our right flank from the sands of the Mojave Desert like an anomalous tumor, canalizing my Platoon into a narrow maneuver corridor several hundred meters wide. The cool temperatures and rare, recent drizzle kept the dust down as my tanks sped through the opening, heading north in a wedge of kinetic, tanned steel. My Platoon was leading the task force on a deliberate attack against an enemy arrayed in defenses just beyond Siberia, an appropriately named expanse of increasingly plentiful wadis, crevasses, and volcanic rock several clicks beyond the Whale Gap. But for now, the sands were flat and inviting, the perfect terrain for our speeding tanks, and we took advantage of it. In those few blissful moments, free of artillery simulators, flashing whoopee lights, and CS gas, I stood (a bit too) high in the hatch of my tank, and, looking back at the spreading array of armored kick-assness racing across the desert behind me, I thought to myself: "Man, this is it, this is what it's all about; I could do this forever—and I damn sure better!"

It wasn't but a few short months later that I found myself tucked away in a small utility room in the Brigade headquarters S1 shop, my only company a surly copy machine and an obnoxious 24-dot matrix printer the size of a Mini Cooper. My too-few years spent in the hatch of a tank had passed, and my days of dashing about the desert (futilely) in search of or defending against the OPFOR were over—at least for now. I had traded in my Tank Commander's .50 caliber for a temperamental computer of suspect reliability, and my other weapons had devolved into spreadsheets, PowerPoint slides, and ponderously plentiful emails.

Apparently I too had devolved, from a once-confident, cocky Platoon leader of lethal men and machines into a mere staff pogue, now responsible for manifest rosters, administrative trackers, and occasional copy-machine kick starting. On my lunch breaks I would often drive by the endless motor pools and look longingly at the rows of orderly gun tubes and wonder when I would be able to ascend the front slope once more and settle into that venerable hatch.

When you're in college, the cadre never tell you about being on staff; it's always flanking maneuvers and L-shaped ambushes, not routing slips, slides, and toner cartridges. It didn't become apparent to me until much later that despite being commissioned as an Armor officer, I would rarely get to feel that comforting tank underfoot. That I would, in fact, spend two to three years on one kind of staff or another for every year I was privileged to lead fellow tankers into combat or in the preparation thereof. I rarely gave a thought that I would have to serve penance in the purgatory of staff again and again as I waited to get called up once more to "The Show."

Unfortunately, perhaps as a symptom of our military culture, it took me far too long to really understand that my role as a leader didn't wane with that assumption of a cubicle on some ignoble staff. There were, in fact, great Soldiers of all ranks deserving of solid leadership and example walking those hallowed halls as well. Staff wasn't a place to simply 'mark time'; it was yet another opportunity to lead Soldiers on a daily basis in an effort to better both them and the organization. The objectives may have changed as had the tools and weapons, but the fundamentals were all still there for the taking, or rather, for the applying. When I finally did grasp this concept, my life on staff changed for the better, and I'd like to think it changed for my subordinates as well. Let's face it: despite all the hype, life on staff is no picnic. But I think in many regards, some of our cultural norms and perceptions about staff have provided us with an excuse to "lead less" while serving in such a capacity, undoubtedly to the detriment of Soldiers, junior leaders, and our own units.

In 2010 I was inspired to write the first edition of The Iron Major Survival Guide while serving as a tactics instructor at the Command and General Staff College (CGSC) at Fort Leavenworth, Kansas. While at the podium there, I often scraped the program of instruction, grabbed a dry-erase marker, and took to the boards in an effort to provide some cautionary tale, anecdote, or otherwise to my class of sixteen eclectic Field Grades. Although not the most gifted or riveting of instructors, I possessed a trove of past professional blights, missteps, and maladies that would serve to source one ad hoc lesson after another that confounded many a day's carefully laid out learning objectives, much to the continued frustration (benefit?) of my captive audience. The Guide was born in an effort to expand the audience beyond the mere confines of my classroom, and convey it across the Army so that more Field Grade officers might learn from my myriad of mistakes and experiences before they assumed those tough Iron Major jobs. Now this newer and more comprehensive edition attempts to take these concepts a few steps further by broadening the audience once again to include staff leaders of all grades, both willing and reluctant ones.

"What qualifies this guy to write such a document?" you may ask. Well, there you catch me at a loss; outside of a few years serving on one staff or another, frankly, not much. I am admittedly an average officer at best, one of the unwashed masses, and wholly unremarkable. There will never be an autographed, star-festooned photo of Dave Dunphy adorning the walls of Marlow & White, and no street or facility named on my behalf—short of a slit trench perhaps. Lamentably, it is only recently in the twilight (demise?) of my career that I've been able to look back over my years of service—some of dubious quality—with a sense of clarity and occasionally, harsh and critical introspection. This death bed perspective, this belated sense of cognizance coupled merely with a desire to see those that follow benefit from my own experiences—both good and bad—is all that I really possess. That, and the motivation to put it into print for the world to see, as if it were my final manifesto. There are certainly far better leaders than I out there qualified to produce such a reference guide as this, but hey, isn't it refreshing to finally see a book on the shelf from the 'regular guy' instead—one that won't bedazzle you with the story of how he made it, but rather why he didn't make it?

It's often hard to discern the contour lines of your map under all those overlays (digital or otherwise). I suppose much of what I've written in *The Iron Major Survival Guide* is common sense; I certainly hope so. Sometimes, however, our environment has a way of swaying us from the course, disrupting our intentions, inducing reticle drift and causing us to lose focus on what's important in life and in our careers as professional military leaders. Hopefully this guide will serve to re-orient and re-boresite us, and ideally, cause us as leaders to pause occasionally for some self-reflection and objective critique, both of which can provide for a re-energized approach to making ourselves and thus our units better.

There are few greater social norms in the Service than our chronic desire to after-action review our individual and collective actions. This candid look at ourselves during combat, training and otherwise, continues to be the single biggest instigator of necessary adaptation and change in this great Army (well, right behind a dwindling budget). If we can consistently apply the same level of humble, honest introspection to ourselves as leaders in a timely enough fashion, imagine the continued and exponential growth we as an organization can experience.

So, why write about Majors? As mentioned earlier, there is potential wisdom within this text that can be applied to all ranks and perhaps even to those outside of a uniform. There is little doubt, however, that there is something special about Majors, as well as the unique challenges—the crucibles—that most of you will imminently face. Let's tally up the numbers. Our 'Five Year and Fly' classmates have long since departed. Endowed with intimidating job titles, enviable facial hair, and the frequent mullet or turret-overhang, many now perch comfortably in the mid and upper echelons of various companies or prestigious corporations. Plenty of the weak-bladdered, chronic underachievers, or persons of ill repute have also left our ranks, sometimes unable to cope with the rigor, responsibility, and character that such service truly demands. There are also those who have served honorably for many years, paid their dues or penance, and for one noble reason or another chose to move on to greener pastures. Many of them continue to serve today, often in uniform-less capacities. And then, among a slimmer company of devoted and seasoned warriors, there are those who have exerted themselves for well beyond a decade on behalf of the nation and its Soldiers; those possessed of the steel, caliber, and character that can perhaps best be described as 'Iron Majors.' Having completed Company command and possibly Intermediate-Level education, they have consciously or unconsciously committed to a career in the Army in some capacity, and now face another battery of tough, key and developmental jobs out there that may arguably be their toughest and most impactful to date. These Iron Majors bear the unforgiving and unrelenting onus of running our Battalions, Brigades, and in most cases, our Divisions. Their performance in this regard will undoubtedly determine their uniformed destiny, for better or for worse. Well, for that choice group of lucky bastards and those like them, this book is for you.

What this Guide strives to be: a "Don't Step in That" compilation of obvious and not-so-obvious precepts that just might enhance one's ability to lead and manage staffs and organizations of all sizes. It may also offer up some career advice, potentially highlighting occupational opportunities and pitfalls. It encompasses the original spirit of the first editions of The Iron Major Survival Guide while attempting to expand upon those lessons to include staff leaders of multiple grades in response to your magnanimous feedback.

There is mention of both science and art within, insights you might not extract from a field manual, SOP, professional journal, or historical reference. Much—if not all—has been discovered, spoken, passed down, or written more eloquently before, perhaps at the end of a command and staff meeting, a farewell remark, a Friday safety brief, a dining-out guest speech, a Company Commander's posted leadership philosophy, or a Platoon Sergeant's introductory pitch to his or her unit. This guide merely serves as a medium to codify some of this wisdom for the benefit of all.

What this Guide tries not to be: the Quintessential Doctrinal Reference. I occasionally attempt to exploit some perceived gaps in our doctrine, but you won't find me quoting a lot of it here. Admittedly, I can barely keep up with the seemingly whimsical and frequent changes in our vast, dynamic dogma. Hopefully the subjects discussed herein are timeless and resistant to the ebb and flow of doctrinal tides (I'm glad to see that the handy term 'supporting effort' is en vogue yet again, but for how long?). Ah, but don't mistake my glib for disdain. Our rapidly changing doctrine is merely reflective of the complex and undulating environments we fight in today, and its continued attempted mastery is an inherent part of our profession—no matter how fickle it may be.

This guide will also not serve as a moral or ethical compass. I won't dwell much or preach upon the Army Values or the importance of ethics in the workplace, etc. I expect that these are given and well-established qualities already, at least from an audience who would take the time to read this book. This guide probably won't help to modify your character in any significant ways; that obligation is purely up to you.

I hope that by reading this, you will end up with far fewer regrets than me. I confess that much of this work was constructed with beer in-hand, so if you feel I have transgressed, that my attempted light-hearted, IPA-supported poking is anything but, or perhaps that I failed to include something of some great importance, please contact me at your earliest convenience. I will attempt to continue to edit, update, and reissue future editions of this book based upon your recommendations, guidance, chastising and feedback. As always, I truly wish you, your families, and your units the very best of luck in all of your respective endeavors.

Best Regards,

David W. Dunphy
Lieutenant Colonel, U.S. Army

Table of Contents

Chapter I: Arithmetic—Attacking Those Decaying Fundamentals 13

- There's no School like the Old School: Know and Demonstrate Military Customs, Courtesies, and Etiquette 14
- Write Good 16
- Brief Well 19
- Slide Well 22
- Proofread, Dam it! 25
- Cross the Line of Departure with Grace 26
- Don't be an Elitist...Ever 30
- Maintain a Fitness Ethos 33
- Email: Combat Multiplier or Lurking Insurgent? 34

Chapter II: The Algebra of Maintaining Good Habits 38

- Run Meetings Well 39
- Manage Your Time 41
- Counsel Your Minions 43
- Plan Sleep 48
- Support the HHC Chain of Command 49
- Master and Exploit Mission Command/Command & Control Systems (and DTMS) 52
- Lost in Space: The Fickle Frontier 54
- Conduct First-Class Ceremonies 57
- Treat Attachments Like Invited Guests 59
- Master Army Property Accountability & Management 62
- The Touchy Taboo of Religion in the Rating Scheme 64

Chapter III: The Trigonometry of Staff Management 67

- RIP a Good One 68
- Manage Your Staff 78
- Lead Your Staff 80
- Develop a Staff Leader Philosophy 81
- Staff Choagie Professional Development (SCPD) 86
- Build Teams Well 91

Write Evaluations, and Write Them Well	96
Develop and Manage a 'System of Systems'	102
Liaison Officers: Required but not Authorized	104
Make Time to Train the Staff on the Military Decision Making Process (MDMP)	106
Those Bastards Down at Company	107
Those Bastards Up at Brigade	110

Chapter IV: The Calculus of Field Gradeship — 112

Iron Relationships or Bust	113
Take Care of the Boss	118
Working with the Boss	119
Pick Your Battles Wisely	122
Learn How to Constructively Advise	123
CAV, Baby!	124
AAR Everything	126
On Occasion, Take a Knee... All of you	131
True Transcendence: the Selfless Team Player	134

Chapter V: The Beer Math of Doctrinal Consumption — 139

Train to Fight; Fight to Train	140
Think Outside the Maneuver Box, but only after You've Learned to Think Inside it	150
Rehearse Well	153
Know the Adapting Enemy better than the S2 (Unless You're the S2)	155
Own Intent	159

Chapter VI: Live to Fight another Day — 164

Don't Estrange Yourself From Your Senior Rater	165
Understand the Army's Career Fields and Functional Areas	167
Stuck in a Purple Haze: Notes from the Joint World	169
Manage Your Own Career	173
Some Parting Deep Thoughts...	175
Keep Things in Perspective	178

Epilogue: Culmination in the Breach — 180

Glossary of Acronyms	192

Chapter I

Arithmetic—Attacking those Decaying Fundamentals

THE INTERESTING THING about the fundamentals is that often enough, the further up in rank we get, the more we inadvertently distance ourselves from them, despite the additional years spent imbued in the military culture. I'm sure this naturally occurs as our priorities evolve, but there are just some things we should never let go of, no matter how trivial we feel they may be. To do so merely portrays an amateurish image to our ever-increasing wards, no matter how many steps of mission analysis we can reel off the tops of our heads. So, for the greater good of all, let's review the bidding on some of those easily perishable basics before some furl-browed Sergeant Major tightens up your shot group for you, in a much more direct and graceless fashion.

There's no School like the Old School: Know and Demonstrate Military Customs, Courtesies, and Etiquette

DON'T SLACK. Don't take advantage of the close proximity you may have with your bosses. Don't become desensitized to senior officers' presence. Stand when the commander enters your office (unless told to do otherwise), and always stand when other subordinates are in the room. If not the Command Sergeant Major, be the person who brings meeting attendees to attention when the commander makes an appearance. Don't drop "sirs" and "ma'ams," don't slack on saluting, and always walk on the left side of a senior officer.

Amateur hour. Nothing says 'Bush League' like an officer caught alone in the open saluting to 'Retreat' before the cannon fires. Know what to do during Reveille, Retreat, To the Color, the National Anthem, etc., indoors and out, in formation and out, and in uniform and out. Don't be the bottom-feeder who knowingly executes a last-second dash for overhead cover in order to avoid rendering courtesies to the Colors when the bugle starts or that cannon fires. There are few greater ways to lose respect.

Civvies count. Be cognizant of your appearance when out of uniform. Hail and farewells, organizational days, and other unit social outlets should not be the time you choose to demonstrate your societal divergence by piling on the nose-rings, gothic make-up, four-day Fu-Manchus, and FTA t-shirts. Invest in a conservative wardrobe for events such as these as well as for government travel, and be aware of what your disheveled or unorthodox appearance conveys to subordinates and the general public through day-to-day interactions on or around base.

Whether in uniform or out, if you wear your sunglasses at night and indoors, then you may suffer from a touch of douchebaggieitis. Unless you just got your eyes dilated at the clinic with the other cotton-tops, take your sunglasses off inside, while conducting key leader or media engagements, or when interacting with foreign nationals.

Dubious practices. Dispose of those dirty habits you picked up at Command and General Staff College. Don't walk & talk, text, or salute on a cell phone. Don't stand around in uniform with your Air Force gloves on in front of Soldiers, and ditch the fluorescent, Hello Kitty book bag. Don't disparage the boss (this includes the Commander in Chief) or gripe and complain in front of subordinates, on social media, or beyond the seclusion of trustworthy closed doors. Keep in mind that everything…and I mean EVERYTHING you post on social media will be seen and used against you for eternity.

The courtesy (and efficiency) of punctuality. The meeting doesn't start at the top of the hour; the meeting starts when the senior ranking chair enters the room and sits down. Don't be the guy who walks in after, frowning at your watch in an effort to convey your self-righteous indignation as to why the meeting started before the posted time. Being on time in our line of work traditionally means being early. A good rule of thumb for appointments and meetings with senior officers: show up ten minutes beforehand for O6-level gatherings, and add at least five minutes for each additional O-grade above six.

Render courtesies from afar. Introduce yourself beforehand when moving to a new unit or position. Sending a letter of introduction to your soon-to-be boss or Brigade Commander shows a touch of class and reach-back to good Army tradition and etiquette. Keep it short, explain your general timeline, and that you look forward to serving 'in any capacity' at that great unit (more on this later).

Guest etiquette. If the boss has you over to the house for dinner or a New Year's Day reception, don't show up in cut-offs and an Iron Maiden t-shirt, don't over-drink or overstay your welcome, and ensure you follow up promptly with a thank-you note to your boss and spouse that briefly conveys your deep appreciation for their vast hospitality, suspect meat loaf, and depleted uranium fruit cake. Add some class by bringing a bottle of wine; but don't show your ass by making it a white zinfandel.

Write Good

MOUTH-BREATHER ALERT. Tired of hearing bosses, instructors, and the Commanding General's SGS complain about your crappy writing? There's a reason for the emphasis, and a direct correlation to your reputation and ability to positively impact your unit. Most of your communication to higher, adjacent units, and subordinates/staff will be through writing, on email or otherwise. You can't escape it. Your capacity—and thus your intelligence—will be measured, and your reputation built upon your ability to effectively communicate above the crayon level through email, memos, situation reports (SITREPs), SOPs, training guidance, policies, 15-6 investigations, Serious Incident Reports (SIRs), etc.

The fundamentals. No one is asking you write like Hemingway, but at a minimum, figure out the following basics of military writing:

- **Less is more.** Dispose of the flowery prose, and avoid the 'atrocity of verbosity'. Your bosses aren't going to reward you based on volume and intimidating five-dollar words. Write plainly and concisely and shun the chronic use of the ARP (Army Run-on Paragraph). Consider a 'Unit Status Report (USR) Litmus Test'

> *Our greater challenge is not in the generation of material, but rather in its trimming into concise, easy, and intuitive reads.*

when composing military documents of any type, from memos to operations orders. Think about the few USR statements you're given to convey your unit's chronic issues once a month to commanders at every echelon, including those at the highest levels of the Army. These meager sentences are all that you have to convince the powers-that-be of your unit's enduring go-to-war problems, and their worthiness of attention. These remarks require brevity, specificity, and potency to be effective or considered. There's a reason why there are so many punishing two-page paper maximum requirements at professional education courses. Our greater challenge is not in the generation of material, but rather in its trimming into concise, easy, and intuitive reads.

- **Punctuation vexation.** Among the many and varied grammatical pet peeves of a flag officer, few will instigate ire more so than an apostrophe faux pas. This includes apostrophes in conjunction with acronyms (e.g., "shoot at those TRP's" or "print up those NCOER's"). Are these meant to be possessive, plural, or are they just contractions?). Know the difference between a colon and a semicolon, and don't use the latter as an excuse to generate ARPs. And don't forget about commas, the most neglected, and, overused, of punctuations.

- **Bullets.** In most cases, it's accepted that bullets need not be complete sentences, but ensure it is written well enough to intuitively stand alone. Be consistent with or without punctuation at the end of the bullets. If you choose to use a period, it most likely indicates that the bullet is, or rather should be, a complete sentence.

- **I.e. verses e.g.** Another popular GO/FO pet peeve (were they ALL English majors?). For most of us, we tend to default to 'i.e.', but I think you may be surprised at the differences between the two, and which one is probably more appropriate in your daily writing (and speech, for that matter).

- **Acronyms and Abbreviations.** If the military didn't use acronyms, we would have to add another hour to the workday. Plus, what better method to hold an entire conversation in front of civilians while maintaining perfect operational security (OPSEC). So, we dig our acronyms, but let's be cautious in their use or misuse. Don't assume the audience of your policy memo or slide deck has any idea of what your abbreviations stand for. Your first use of an acronym (ACRNYM) or abbreviation (ABB.) should spell it out in its entirety, followed by the actual acronym in parenthesis (PNTHS). You now have carte blanche to use the ACRNYM by itself for the remainder of the document.

[Note: overachieving proofreaders delight. I'm sure after that last section, there will be a few cats out there who will spend quite a bit of time drilling down on this document with enthusiastic scrutiny, zealously pointing out every hypocritical writing faux pas. When you get it all figured out, feel free to forward your analysis to me. Candid feedback is a cherished, inexpensive gift; if only my editors operated in such a fashion...]

Brief Well

LOUNGING AROUND BEHIND a "Command Post of the Future" (CPOF) screen in your Spiderman Underoos with headphones and the highlighter tool in hand does not necessarily a good briefer make. As the ease and cost savings of technological 'benefits' pervade our cultural norms, things like video teleconferences and Situational Awareness system updates have become commonplace in our profession. Although there is some art associated with presenting in these kinds of mediums, we shouldn't let go of fundamental briefing skills simply because we've rid ourselves of Combined Arms Services Staff School (more's the pity). Whether you like it or not, you will spend a good portion of your professional career standing up in front of a room or tent full of senior-ranking people attempting to conduct one type of briefing or another. Like a shy, bumbling high school English class speech, few things will earn you a post-event wedgie in the bathroom like a botched brief to the Command. Let's review some of the often-discarded basics:

Work to develop steady, crutch–free speech that inspires confidence in your audience by rehearsing ahead of time and owning the material through its complete understanding.

Know the terrain. If you are unfamiliar with the preferred briefing techniques of the senior officer in the room, always default to a conservative stance—literally. Assume a modified or 'loose' position of attention at an open vantage point on the stage left or right side of the screen or dry-erase board, and stand relatively still. You should have the ability to observe both the screen and the audience without exhibiting excessive movement. Clasping (not wringing) your hands tends to be permissible by most. Minimize the weatherman's excessive hand and arm signals, unless attempting to direct attention to an item on the screen by using the hand closest to the screen, so as to not bring your shoulder or back to the audience. No pacing around like a hungry zoo leopard, and no fat, greasy paws on the screen. Always use a pointer and never talk to the screen. Open your brief with a quick welcome, introduction, and salutation, and maintain eye contact with the audience.

Pointer etiquette. When using a wooden pointer, refrain from demonstrating your fencing skills or working on your backswing. If you're not pointing at the screen, keep the tip down and remain still. If you are using a laser pointer, exercise some muzzle control and be conservative with your movement and its use—no concert special effects support here. Be cognizant of laser fratricide & reflection, especially if briefing from a flat-screen or plasma TV. Some laser pointers are downright obnoxious, so save those super-nova green monsters for the convoy night live-fire, and choose from a moderate, less distracting arsenal, still viewable in the daylight.

Table drop etiquette. Before kicking off the brief, spend a minute orienting your audience to any handouts that you might have placed in front of them such as slide decks, back-ups, spreadsheets, etc., especially if there are multiple items. If you are referring to any fixed maps or detailed sketches in the room or on a slide, pause to provide a quick orientation to those as well.

Find a happy medium. When briefing from slides, there are typically two extremes to avoid: (a) once the slide appears, you simply let the audience read it in awkward silence before making the somewhat arbitrary decision to advance to the next slide, or (b) you start with the slide title and read every bullet, verbatim, from top to bottom—much to the horror of your captive audience. There's probably a happy medium in between, but conduct some reconnaissance to determine what technique the boss tends to lean toward. I recommend selecting a few important points to discuss per slide, and do so by pointing to the bullet, then facing the audience squarely and expounding upon the issue.

Allow time for the audience to absorb the remainder before segueing to your next slide. And yes, don't forget to segue; it provides the audience necessary context and connectivity from one slide to the next.

The 'ugh' in 'uhs.' Everyone has a crutch or stall word readily available to tidy up those seemingly uncomfortable spaces between words, bullets, or slides. 'Uhs' and 'ums' are tolerable, but only in their most frugal employment. As a Small Group Instructor (SGI), I once counted over 115 'uhs' in a student's five-minute terrain and weather analysis brief. With every subsequent utterance, the word (or lack thereof) drilled irritatingly deeper into my cranium at such a point that I was unable to focus on the content of his brief in any real capacity. When I told the officer how many 'uhs' were uttered, he was incredulous, but the rest of the tortured audience quickly stepped in to voice their own frustrations. Speaking in public can be a nerve-racking event for many of us, especially before a room full of peers and higher ranking leaders. Work to develop steady, crutch-free speech that inspires confidence in your audience by rehearsing ahead of time and owning the material through its complete understanding. Side note—I would refrain from picturing your audience in the buff, however—unless you're into that sort of thing, you freaky-freak.

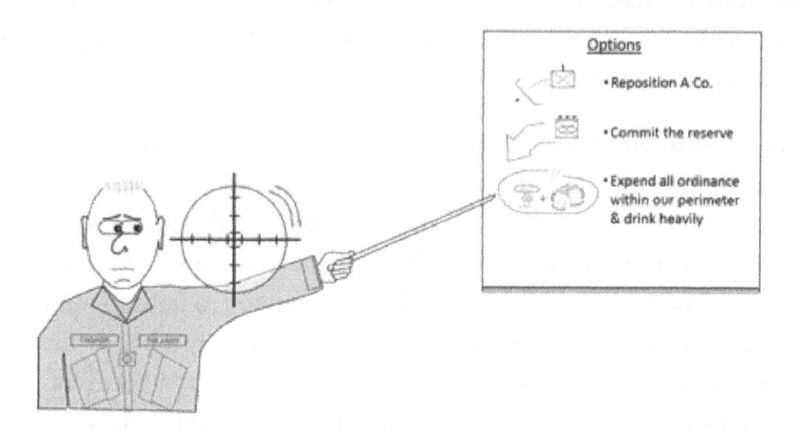

Man down. When the cross-hairs site in center of mass and the sharpshooting begins, don't be in a rush to answer questions based on only your partial or absent knowledge of the subject. Don't be afraid to say "I don't know, but I'll get back to you with an answer ASAP." Have a helper off-stage capture the question, issue or RFI, and review the due-outs at the end of your presentation if time permits. If this expression becomes endemic during your briefing, however, I would submit you might want to master the subject more thoroughly before presenting next time.

Just don't bore me. Most of what's been discussed thus far tend to be more scientific briefing aspects. Some art comes with having the ability to make a quick determination as to the audience or chairman's knowledge of the subject, and then modify the pace of the brief accordingly. This is especially true of executive-level or deskside briefings, where time is even more valued, and your loquaciousness isn't. Bottom line: if the audience looks uninterested, speed it up.

Slide Well

I ALWAYS FIND IT AMUSING when some crusty cat in the back of the room says, "I hate slides, whatever happened to good old fashioned butcher block?" Well, there was nothing fun or efficient about butcher block save the scented markers and their collateral benefits. And overhead VGTs? Forget about it. PowerPoint slides are here to stay, and are so deeply entrenched in our culture that we have no choice but to figure them out. That being said, let's not get too carried away here. There's a reason why they have such a bad reputation. If mismanaged, they are, arguably, the biggest time-killers in the history of military staff operations (right after SimCity).

Keep it simple. Again, like briefing an audience, one must know the terrain before embarking upon successful slide creation. The rule of thumb to consider is 'good is good enough.' When PowerPoint first started to permeate our ranks, we were just happy to be able to quickly put big words on big screens for the benefit of big groups. As our culture became more and more familiar with this medium, a type of arms race began where complex diagrams, clip art, pictures, 3D effects, shadowing, hyperlinks, and finally animation became commonplace—and, to the horror of staff officers everywhere, even expected. These things are often cute, but rarely necessary in order to convey information or prompt an educated decision. Manage expectations early and accordingly, as the time required to prepare such gaudy 'money slides' is an exponentially increasing function, and is most likely detracting from other, more important efforts.

It's not about showing off all of your cool unit crests and Microsoft savvies as much as it is about conveying information in the most effective, concise, time–preserving, and easily deliverable method possible.

Size matters. The more complex your slides are, the larger the presentation becomes, and the more bandwidth is required to transmit them. This may not be a limitation in the luxuries of garrison fiber–optic, but pushing large, memory–laden slide decks over SATCOM or line–of–sight devices between echelons in the field can be problematic. Bottom line: maintain an appetite suppressant when creating visual 'cornucopias' for the audience. It's not about showing off all of your cool unit crests and Microsoft savvies as much as it is about conveying information in the most effective, concise, time–preserving, and easily deliverable method possible.

Don't overwhelm the audience. Digital paper is free (unless you hit 'print') so don't feel you need to compress 250 words per slide. Ensure your font size can be read from the back of the room, and try to limit points or bullets to a maximum of about 6–9 per slide. Simple and intuitive 'quad' charts are also handy, just don't let the font size shrink too much. Highlight important information with a bold font or a consistent, different color to provide the briefer a quick crutch or prompt to refer to while briefing. Grey-out stale but required milestones or information, and utilize a stark blue for new data or updates. Don't get too colorful with the fonts, as this will only prove distracting, and perhaps beg questions about your various color schemes. Shy away from wingdings and fonts like Times New Roman, as these too can be hard to discern from afar. Arial and Arial-like fonts should be the preferred choice for most presentations. But whichever you select, ensure every one of your slides, to include headers, footers, and page numbers, is consistent. If you are developing a stand-alone presentation, then exploit the 'notes' pages for more in-depth or supporting information.

Slide admin. Regardless of whom you are briefing and in what forum, ensure you include the requisite classifications on your products. If your entire slide deck is 'UNCLASSIFIED//FOUO', your slides need to be labeled as such, and there is no need to put a '(U)' before every bullet. However, if your slides are of a higher-level classification, you have a bit more work to do. Don't screw this up, as heads may very well roll— most likely yours, not to mention the embarrassment of being called out in public by some pale-skinned intel guy ("you really need two forward slashes in-between there, champ—squeak"). For long lists or slides with more than five or six bullets, consider using numbers or letters instead of traditional bullet icons or dashes. This will allow a member of the audience to quickly refer to a point on your slide verses the awkward and time consuming comments like "your bullet about half-way down the page may not be a valid assumption." And don't forget page numbers. They're necessary reference points for meetings and presentations of any type, especially if slides are used in video-teleconferences (VTCs) or DCO/DCS meetings.

Don't shirk, but manage expectations. Keeping slides simple is no excuse to half-ass the presentation. Ensure your slide deck is organized, grammatically sound, cohesive, consistent, and concise. A glaring typo or misspelling on a slide can be an embarrassing detractor from your pitch, so spend less time making objects and words fly around your screen in three dimensions, and more time proofreading and ensuring a quality product, no matter how Spartan. However, if you tell your subordinates to hastily throw some slides together and brief you in an hour or two, adjust your quality expectations accordingly.

Proofread, Dam it!

CAUTION: SPELL CHECK ≠ PROOFREAD. Don't simply rely on subscripted squiggly red lines to highlight your prose inadequacies, and don't bear the burden of the corrections effort.

Defend in depth. As a staff leader, you will or should not be the first echelon in the QA/QC process. And, as an operations or executive officer, you should be the last (not only) line of defense for any and all products going to and through the commander and beyond.

Systemic QA/QC. You may be amazed at how poorly some of your 'college graduate' staff officers and NCOs write (evaluations, awards, intelligence summaries, operations orders, unit Facebook pages, etc.). Develop quality control systems on your staff or in your section to ensure a poor product does not end up on your desk without someone other than the writer having proofed it first (hint—find the English or History major on staff and stick him/her with the occasional duty; they may not appreciate it, but what a great way to punish those slacker liberal arts majors for catching all those happy hours and keggers while you were sweating over some design project). Otherwise, you will spend a good portion of every day 'grading' and rewriting subordinate products—another huge time killer.

Be demanding. Be meticulous early on with staff products, and pay attention to detail (format, commas, grammar, cover sheets, routing slips, page numbers, headers, etc.). Be ruthless up front with rewrites & re-dos, and the staff will get the message quickly. Based on your assessment, you may want to consider working writing skills into your staff's professional development program, or assigning homework and articles to further their skills (or lack thereof). Make it painful for them, and eventually they will make it painless for you.

"Edit this please, Sir." Don't let your boss be your designated proofreader. Have a peer—or even a subordinate—proofread your products before they go forward.

Cross the Line of Departure with Grace

AS PREVIOUSLY MENTIONED, reaching out to your future bosses through correspondence before your arrival is a cherished but dying courtesy. It is also one that, if done correctly, can set you off on the right foot immediately. A reputation that will span a decade can be made in mere minutes.

The gamut of correspondence. There is no doubt that such an action is a classy (not cheesy) move, and the spectrum can range from more informally emailed notes to crisp, clean, snail-mailed letters on decent stationary. The former may be more appropriate for your immediate boss, while the latter is perhaps better suited for his or her boss (your senior rater).

Refrain from telling your future bosses where you think your exemplary talents should best be applied.

Don't just call. I didn't include phone calls as a viable substitution for an email or letter for a reason...it's lazy and makes a poor first impression. A phone call is certainly a necessary requirement or follow-up, but perhaps better applied toward the unit adjutant, sponsor, or outgoing trooper verses one's direct supervisor or, heaven forbid, senior rater.

Recon. I recommend researching through their respective adjutants as to which technique is appropriate for your bosses. Don't be surprised when they respond to your inquires with "no one sends those; don't worry about it." Politely provide that cat with your contact information, and then get to typing.

Caution. Take note, however, that any correspondence between you and your future career shapers—regardless of the medium—should be cheese-free and flawlessly written. Also refrain from telling your future bosses where you think your exemplary talents should best be applied. For example, "I would be a great S4;" "please don't put me in the 3-shop;" "I think I have the staff thing down and should skip straight to Company command now;" or "I'll only need to serve a few short months as a Battalion S3 before I'm ready to assume the role as your Brigade operations officer." Such notes somehow tend to find themselves circulated amongst the command for all to view and deride prior to your glamorous arrival. A couple of (relatively sound) letters of introduction examples are provided below for your consumption.

> Sir,
>
> Greetings, I'm LTC Dave Dunphy, and I am on orders to the Joint Force Headquarters J3 directorate with an arrival as early as this spring. If all goes well over the remainder of the week, I'll graduate Joint & Combined Warfare School (JPMEII) here in Norfolk on Friday, 22 March. With your concurrence, I would like to take a few days for some leave and house-hunting permissive TDY in the local area. I plan on signing in at the Headquarters on or about Tuesday, 7 April. I've linked up with my sponsor, LTC Jim Smith, who has already proven helpful during my transition. I am ready to begin work within the organization, and I look forward to meeting you and the rest of the J3 team shortly. My bio and ORB are attached for your casual perusal, disposal, or occasional ridicule.
>
> Best Regards,
>
> LTC David Dunphy
> Armor, U.S. Army

An Informal (emailed) Letter of Introduction

A snail-mailed, old school letter might look something like the following, but I recommend double-checking your personal correspondence format by accessing the latest version of Army Regulation (AR) 25-50. I also submit that this medium requires a bit more formality than my last example (I used the previous note—almost verbatim—at my last unit, and my boss replied with "looking forward to meeting you, smart-ass." Thankfully, he was a man of some humorous capacity, and I wasn't (immediately) relegated to bathroom OIC duties). Remember, this isn't a forum for you to push your personal ambitions or agenda, it's simply a time-honored tradition based on common courtesy to convey your willingness to serve in the unit, and your gratitude for being given the opportunity to do so. Don't forget, the letterhead is typically only used for correspondence outside of your current unit.

Formal Letter of Introduction

DEPARTMENT OF THE ARMY
(OLD) ORGANIZATION NAME/TITLE
STREET ADDRESS
CITY, STATE 12345-0000

April 1, 2015

Colonel Ima F. Tyrant
301 Hell on Wheels Road
Fort Carson, CO 80923

Dear Colonel Tyrant:

    Greetings, I am Major David Dunphy, an Armor officer on orders to the Spartan Brigade with an arrival date no later than 1 July. If all goes well over the remainder of the week, I'll graduate Command & General Staff College at the end of it, and proceed to Colorado shortly thereafter. I plan on signing in at the Division personnel center on or about Tuesday, 20 June. I am in contact with your S1 and have linked up with my sponsor, Major Jim Smith, who has already proven helpful during my transition. I'm eager to begin work within the brigade, and I look forward to meeting the team and serving in any capacity I can. My biography and ORB are attached for your reference.

                               Sincerely,

                               David W. Dunphy
                               Major, Armor

Enclosures

Don't be an Elitist...Ever

DURING THE LATE NINETIES, I served as the Brigade Commander's TAC battle Captain during a trying ten-day Warfighter exercise. We deployed to the field and set up all of the Brigade's command & control nodes in support of this less than cherished event. It was my first day serving in this capacity, hovering around a large map board in a stuffy SICUP shelter where I gained a new appreciation for just how painful life on staff can really be. The Brigade Commander possessed a rare talent of being able to weave together entire sentences composed of nothing but caustic four-letter words, and he pummeled me with them in such consistent volume and vehemence, it was worthy of a Drill Sergeant with a case of the clap. It was not the kind of exposure I ever hoped to have with my senior rater, and after 18 hours of continuous beat down, I was humbled and humiliated to say the least.

Around midnight, he hoarsely barked at me in the standard gruff fashion, "Dunphy, put your kit on, go find my tent, and fill up my thermos from the coffee pot in there—this stuff here is God-awful, 30-weight motor oil." I didn't even have it in me to ask him where it was when I stepped from the tent into the dark of night, blinking hard to acclimate to the pitch of the Fort Stewart forest. I started off in search of allegedly better coffee, cursing under my breath, and slowly stumbling in the general direction I had witnessed some other structures popping up earlier that day. The vitriol thoughts bouncing around in my head slowly took physical form in a venomous low mutter as I blindly felt my way through the thickets, where at one time I simultaneously hyperextending my right knee in a ditch while inadvertently jabbing my eye on a low hanging branch (we didn't use eye pro back then). 'This is straight up crap', I thought, as I limped along bleary-eyed. 'That guy has been riding my ass all damn day, and now I'm out here stumbling around in the woods fetching coffee?', 'I don't remember that being in my job description, I'm a Captain for crying out loud, not some errand boy/gopher, WTF (we didn't have that acronym back then either, but you get the picture)!'

> Somehow, through all of the cursing and rancor, I found the Old Man's tent, and made my way inside to find a tranquil setting like none I could have ever imagined in a field exercise. In the cozy lamplight, the cot looked comfy and inviting covered with a friendly comforter and down pillow. A state-of-the-art coffee maker sat poised on a makeshift nightstand with some effervescent, foofy smelling stuff swirling about within. I unscrewed the thermos cap, reached down and flipped the nob, only to see the coffee squirt out with unpredicted enthusiasm, where, much to my horror, it splashed down upon a set of equally comfy-looking slippers on the tarped floor beneath. "Damn it all to hell!" I shouted out loud, as I frantically patted at the footgear with my sleeve, 'I'm going to hear about this too,' I thought, 'how will I ever make it to Company command after this debacle of an exercise?'
>
> After another treacherous trek through the woods, I made it back to the TAC where the commander eagerly scarfed up some of his 'French pumpkin bisque' Joe, or whatever the hell it was. I withdrew into a dark corner of the tent, where, obscured slightly by map boards and radios, I quietly sulked while awaiting the next contact report, ass chewing, or asinine midnight errand. Thankfully, I never heard back on those damn coffee-soaked slippers.

You are not authorized an aide de camp (at least most of you—for now). Just because you're the Battalion XO doesn't mean there is some Soldier dedicated to loading your duffle on a flatbed truck, setting up your cot, or getting you a cup of coffee.

Why should you get the TMP van to go to the range and qualify? It may be a matter of time management in select circumstances, but if not, put your ego aside and get on the bus or 5-ton with the rest of the staff and HHC troops. Besides, that TMP belongs to the commander and Sergeant Major, and you'll probably Q2 anyway.

Don't get the boo-boo lip when there's no marked chair for you under the awning at a unit Change of Command ceremony. Chances are it wasn't a deliberate slight. Refrain from griping when there's designated parking for that remfy space ops guy, but none for you. Go park in the Back 40 with the rest of the staff and step it out, chubs. How else will you break a thousand steps on your glam fitness tracker anyway?

There are no such things as executive fitness tests, weigh–ins, or tape–tests.

Don't make it a habit of cutting to the front of the urinalysis line every time because you think your rank or position warrants it. What you are really saying to those around you is 'I'm more important than you, and so is my time.' There may be moments when you need to expedite things for a higher cause, but if not, wait there patiently with the rest of those troopers; you may be surprised at how much you can learn from a couple of E4s while standing in line for 15 minutes. Plus, with all of those goofy plans and taskers you've been issuing out of your shop lately, there's a good chance you'll piss hot anyway. Chug some more water, visualize a babbling brook, and enjoy the tranquility of that stress–free slow boat to the John.

There are no such things as executive fitness tests, weigh–ins, or tape–tests. Unless you are wearing a pointy, star–shaped thingy or three on the center of your chest, get your sagging can in line with the rest of those troopers at the next PT test and weigh–in. Your position as S3 or XO doesn't exempt you from the same standards of fitness and fitness evaluation that the rest of your staff or unit has to abide by.

Maintain a Fitness Ethos

DURING A RECENT TOWN HALL MEETING, our commanding General spent a few minutes emphasizing the importance of physical training in the unit. He paraphrased a speech by the Chairman, who stated quite simply that when a unit gets the call to deploy into harm's way, the Army can do many things to rapidly train, equip, and prepare it for combat, but no matter how urgent the call, there are two things you can't do overnight: build teams and get fit. These items require prior investment and constant emphasis to ensure a unit is prepared to fight, especially in today's complex, demanding, and pervasive tinderboxes.

Non–negotiable. Make physical training sacred, both in and out of theater. Don't be the guy who recommends cutting out PT to apply an hour or two to another training event or slide prep for a meeting. You have a lot on your plate, and another hour added to the workday sure would be tempting and convenient, but don't fall into the trap. If it comes to that, then tack on an hour at the end of the day verses trimming out PT.

Rules–of–thumb. You don't have to be the fastest and strongest guy on staff or in the Battalion, but both need to see you doing hard PT daily. Never hang out behind your computer five or ten minutes into the PT session. Get out before the Flag goes up, and make your staff do hard, visible PT, too. A short stretch–ex after Reveille is a good 'bubble–leveler/ staff huddle' before the day, but don't let it overtake your PT session—no pen and paper, and definitely no cell phones or Blackberries. You can pursue more decentralized staff or section PT when you trust that they got the message and your intent for quality training. Give your staff officers the opportunity to conduct PT with their respective sections routinely as well, and don't disregard HHC PT initiatives.

Hercules, Hercules. Whatever you do, don't show up to

your unit fat. You'll instantly lose credibility, and it's a tough hole to dig yourself out of. Don't get fat while in the unit, or while deployed. If you are a borderline turret plug, then drop the chubby snacks, tanker pills, and other Air Force vegetables; grab some carrot sticks and bust any preconceived notion that you're just 'big boned.' It's worth repeating here that if you need to get taped, then get taped by the HHC First Sergeant like everyone else. No closed-door tape sessions or one-on-one PT tests with the operations Sergeant Major. Soldiers will only assume the worst.

Email: Combat Multiplier or Lurking Insurgent?

IT WAS APPROACHING 1600 on a balmy Friday at Fort Stewart, GA, and the number of Soldiers anxiously eyeballing the parking lot grew steadily in anticipation of closeout formation and the weekend safety brief. As the Platoon Sergeants gathered the men—most envisioning a jovial happy hour or fun-filled family night—I received that fateful COB phone call that all tankers dread: an engine (pack) had just arrived for one of my Company's bent tanks, and we had to get it in today so the unit didn't suck down another weekend of having a downed pacing item with parts on hand.

I delivered the news to 2nd Platoon, and the sullen tank crew, mechanics, and chain of command lumbered down to the motor pool—with all notions of a Savannah Friday night on the town dashed. Nothing with tanks is easy (except killing bad guys), and in true form, this pack proved to be a stubborn and time-consuming job. At 2030 though, we finally hooked up the fuel lines and prepared to fire up the engine. With an audience eager for the impending, comforting whine of an operational tank turbine, the driver slipped into the hatch and hit the start button. We were greeted, however, with the continued and disappointing sounds of silence. After another 30 minutes of troubleshooting, the mechanics declared that the rebuilt engine delivered that late afternoon was, in fact, dead and broken (but nicely painted). Dejectedly, we threw in the towel, and the despondent Soldiers finally made it out to the darkened, mostly empty parking lot.

I was so incensed at this colossal waste of time, that I went back to the office to compose a vicious and sarcastic email decrying the night's events, and ridiculing the logistical train and the quality of the shabby, crappy engine rebuilds we continued to receive. I didn't hesitate when I angrily hit 'send' in a dramatic and self-righteous fashion, and hoped the Battalion XO (long gone by now) would somehow set the world straight so we wouldn't repeat such an atrocity again. By noon on the following Monday, however, I had an inbox filled with scathing emails from the DISCOM Commander, Brigade XO, and plenty of other senior ranking folks who didn't take too well to my log-bashing tirade, as the Battalion XO had simply forwarded my email to higher HQ, where it made its way quickly around the Division. As a result, the parts bins for Alpha Company (Wildbunch!) would continue to dry up steadily over the next several days, and those drawn MRE lot numbers would get older and older.

Live in infamy. Talented leader's careers have ended prematurely because of email or social media blunders. If you don't want to see what you wrote posted in The New York Times or sitting in the inbox of the Commanding General, then don't put it in an email (this includes Facebook and Twitter). As you know, your original email may be attached to a long line of forwards, etc. Once you hit 'send', your words will live in infamy, whether you want them to or not. Remember, emails may be personal but they are never private; adjust accordingly.

> *If you don't want to see what you wrote posted in The New York Times or sitting in the inbox of the Commanding General, then don't put it in an email.*

Think twice, send once. Refrain from writing and sending emails while angry (or after a few drinks, for that matter). If you can afford the time, take a break for a bit or sleep on it before sending. You will be amazed at how differently and less contentious the issue looks like the next morning or even after an hour. When possible, have an objective peer read it before sending. If you demand instant satisfaction, then pick up the phone or, heaven forbid, get off your can and go visit the object of your wrath.

Forwarding leadership. Apply some analysis to emails. Don't manage or 'lead' your staff by forwarding higher headquarters' or the boss's orders. Make them your own. An "FYI" on a forwarded formation time change is acceptable, but when the boss writes you and says "I'm tired of units submitting their Green-2 reports late," don't simply forward to Company Commanders and preface with, "please note the Battalion Commander's comments below."

Email Creep. Don't let email become your sole or dominate managing style (note the deliberate absence of the term 'leadership style'; you can half-assly manage by email, but you can never lead through it). Pick up the phone, get out of the office and go find a subordinate; talk face-to-face whenever you can. You will find that such personal interactions will add depth to an otherwise bland day, while granting you critical time to work on those neglected people skills. It will also enhance your situational understanding with the added benefit of giving you ample opportunities to get to know your minions that much more.

Action sent is rarely action received. If you send an important, time-sensitive email or one that requires due-outs, ensure you follow up with a call to the recipients directing them to read it. Staff Duty can help with this.

Archive everything. Save every email you send or that has been sent to you (except the risqué stuff). An email historical archive may very well save your hide someday. Organize permanent, archived folders for deleted items, sent items, special topics, etc. Consider creating a special folder for 'Boss Emails' that will allow you to review, in one location, all of the taskings, GFIs, or work priorities your boss has sent you.

Tracking like a SCUD. Some bosses want you to acknowledge their emails, some don't. It's good to send back a quick "roger, Sir", when given specific tasks, but it's perhaps a bit unsuitable or even cumbersome for an "FYSA" kind of note. Include this as part of your 'boss IPB', or default to a conservative stance until told to do otherwise.

'Reply to y'all'. It can be handy at times, but it usually just fills up people's inboxes with unnecessary battlefield clutter, or gets people in trouble. Be selective and careful in its use, and figure out who's on the 'to' and 'cc' line before hitting send.

Chapter II

The Algebra of Maintaining Good Habits

REGARDLESS OF WHERE you might find yourself, this chapter is devoted to those sound practices that may carry the day in any job or on any staff. Their healthy employment can very well be the deciding factor between a struggling staff/staff section and a successful one.

Run Meetings Well

LIMIT MEETINGS and attendees when possible. Remember, the more people you bring into your meetings means fewer things are getting done or are being supervised concurrently throughout the unit.

Always have a purpose and agenda. If not, your meetings may wander aimlessly, tying up valuable time. Ask yourself, "Why are we having this meeting?" Is it simply to provide an information brief or are you soliciting guidance or decisions from the commander? If it's the latter, ensure you prepare the boss ahead of time instead of springing the blank 'Commander's Guidance?' slide at the end of the brief and putting him or her on the spot.

Keep meetings short. Attention span and bladder volume are inversely proportional to each other; the former will decrease exponentially after 50 minutes (think ILE classes). If briefs were supposed to be long, we'd have called them "longs." Train staff members to be prepared to brief when it's their turn. Wargame or hand-mike rules in effect: when it's your turn, push to talk, not to think; know what you want to say, speak confidently and quickly, then hand the mike to the next guy and sit down.

Be predictable. Schedule routine times of the day/week for meetings (training, Command and Staff, resource, maintenance, Battle Update Briefs, Commander's Assessments, targeting, working groups, Logistics & Maintenance, etc.). Obviously, while deployed don't allow meeting attendees to set patterns if patrols are required to shuttle participants. Combine topics when possible to limit the number of meetings as well. Outside of quick huddles or staff stand-ups, avoid scheduling routine meetings on Mondays and Fridays, as the holiday schedule will predictably disrupt those battle rhythm events.

Plan for the inevitable hasty meeting. Establish a "hey you" meeting window regardless if you fill it or not for hasty IPRs, Unit Status Report

(USR) updates, leader huddles, etc. (i.e. Tue/Thu 1130–1230; lunchtime meetings, although painful, limit training schedule disruptions).

Prep the objective. Send a read-ahead when possible so the audience is smart before they sit down. It will save time, and often prevent embarrassment or dime-dropping. But know that you're not doing any favors by sending the boss or anyone else a read-ahead at 1700 the night before a 0900 meeting; you just added another half-hour to his or her work day by floating that unforecasted, last-minute turd. Give folks at least a full workday to digest a slide deck, or you will most likely face an ignorant meeting chairman, among others. As part of that read-ahead, be upfront with the purpose of the brief to better queue up timely guidance or a decision. Another good rule of thumb is to add a day for every star the senior ranking officer sports, e.g., if briefing the Division Commander, attempt to have that read-ahead in at least two days before the event.

Actually read the read-ahead. If you're the boss or chairing the meeting, review the slides or read-ahead provided before the event. Don't wait until you're sitting at the head of the table during the meeting to push back, stare at the ceiling, and conjecture; you are just wasting everyone's time.

Working and planning groups. Such groups aren't your typical meetings—or rather, they shouldn't be. To manage expectations and to facilitate productivity, slot at least 90 minutes for these venues. To host an effective working or planning group meeting, the participants—all of whom, ideally, are consistent and informed members of the group—will probably need to prepare products or conduct research beforehand. Executing these meetings without prior preparation almost guarantees their degradation into time-consuming information briefs or lectures, as the audience has to be caught up or coached onto the subject matter yet again. These sessions should have an articulated agenda that includes the tasks to be accomplished and why (purpose), along with clearly defined inputs (those things that must be completed ahead of time and brought to the table) and corresponding, codified outputs. Few working or planning groups are effective without some lively, guided, and informed dialogue, a few scribbled-on dry-erase boards, and a stout read-ahead package.

Manage Your Time

YOU HAVE PROBABLY HEARD this throughout your college years and career ad nauseam, but it's only going to get worse from here on out. Establish a nested unit battle rhythm, both in garrison and while deployed. It needs to be nested with subordinate unit events, and certainly higher's. Limit variance to provide predictability for all, and don't feel obligated to fill up those rare, tantalizing white spaces on the calendar. You'll need the flexible room.

The secret of the Grail. The key to managing your own time is to learn to properly manage your subordinate's time and priorities of work, without necessarily micromanaging them (i.e., delegation). In order to master an artful balance, I strongly encourage you to read and apply the genius article, "Management Time: Who's got the Monkey?" by William Oncken Jr. and Donald L. Wass, first posted in the Harvard Business Review in 1974. When you're done with this timeless piece, assign it to your minions for homework and internalization, then conduct a brief professional development huddle on its implications so you can all adjust your expectations and fight from the same common operational picture (COP).

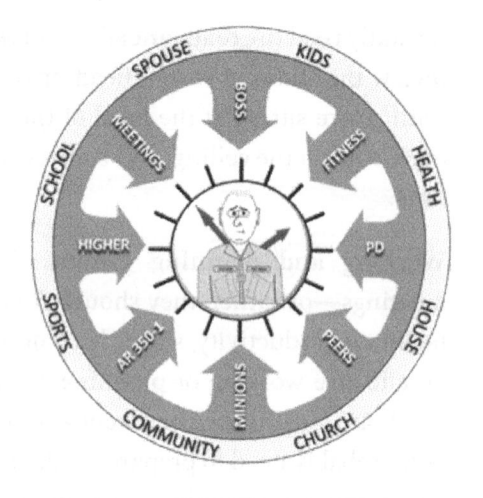

Build routines. Establish your own predictable personal battle rhythm, both in and out of theater. Share your calendar with subordinates so they can see good opportunities to approach you or backbrief you on their work. You may also want to consider setting up reccurring weekly or bi-weekly staff section updates, where you can sit down with a single section leader at a time. In such a forum, he or she can review their priorities of work with you or seek additional guidance in a more intimate, routine, and wholly predictable setting. You will also find that this will mitigate those often necessary but distracting subordinate drive-bys (drive-bies?) over the course of the day, since they now have an opportunity to brief you on their products on a regular basis. Ensure they provide you with a read-ahead the day prior if the subjects warrant it, especially if they are seeking a decision or additional guidance.

> *If you can't get out of the office most nights by 1800, then you are doing a poor job of time and task management, and delegation.*

Build flexibility. Set aside time on your calendar for staff product QA/QC, and build in buffer time for 'firefighting' and the friction of war. It would be nice if the boss approached you at the same brief, predictable timeframe every day to review his priorities of work, but let's face it, that's a pipe dream. He is more likely to send you a dozen emails over the course of the day, leave Post-Its and notes on your desk and monitor when you're at meetings, and conduct 'drive by shootings' when you're in your office. So, when the boss strolls in on Monday morning with a note pad full of GFIs and things he wants you to get done or check up on, your carefully planned daily schedule and priorities just got bumped or dumped. The boss can—and in many cases will—be the biggest contributor to your inability to manage your time, but hey, it's the nature of the beast, and he's the boss. Deal with it by building in the time for it.

The day your boss stops 'bothering' you with such things is the day you have become irrelevant to both him and the organization.

Family matters. Make family a priority for you and your subordinates. Take time off for your kid's soccer game on a Tuesday afternoon. Allow and encourage your subordinates to do the same, but don't get taken advantage of. If you can't habitually strike that balance between work and family, someday your support structure may very well fold up on you.

Shoot the bull. Build time within your daily schedule to rub elbows with your subordinates and Soldiers. On many days, you may be able to afford little more than a brief 15-minute stroll through the motorpool or, occasionally, a Company command post. There's little doubt, however, that you'll find these few minutes will end up being some of the best and most gratifying of the day. Don't overstay your welcome as your presence takes them away from their own priorities of work.

Go home. Don't stay deployed when you're not deployed. If you work until 2100 nightly in garrison, many of your subordinates will feel obligated to do the same, whether they have work to do or not. You will all be miserable as a result, and so will your families. If you can't get out of the office most nights by 1800, then you are doing a poor job of time and task management, and delegation. It is not a badge of honor to work 80 hours a week, but rather an indicator of some suspect or dubious managerial habits.

Counsel Your Minions

Lip service. We all preach to good counseling, but it's still amazing that at every echelon I've taught, the majority of officers continue to state that throughout most of their careers, they received (and admittedly gave) crappy counseling, if at all. In class or behind the pulpit, it's easy to speak to what right looks like, but in units, it still seems to be the first item (right after your officer professional development (OPD) program) that falls off of the schedule when the calendar gets busy, which, for most of us, is always. Counseling is not hard, but it is time consuming and requires preparation. It is, however, one of the most underrated (and underutilized) combat-multipliers you have in your leadership arsenal. An investment in good counseling will pay off exponentially in a myriad of tangible and intangible ways.

Schedule it. Give advance notice to subordinates, put it on the calendar, and task them to come prepared to talk about the good, the bad, and the ugly. Provide them focus for the session ahead of time. Task them to provide you feedback as well: "S4, what do I need to do or change to better enable you to do your job?" I've found that employing a 'counseling worksheet' filled out by the trooper ahead of time serves as a great agenda and forces subordinates to conduct some self-assessment before the session. Consider the example worksheet questions below:

- What are your professional goals while assigned to the unit?
- What are your personal goals while assigned to unit?
- What is your next major career milestone?
- What do you consider to be your strengths as they relate to your position or the Military Profession in general?
- What do you consider to be your weaknesses in relation to the same?
- What tasks (from your section) do you feel are currently being done well?
- What tasks do you feel need improvement?
- What can I do better or differently that would enable you to do your job better?
- If given the chance, what would you change about our staff/staff section or the unit as a whole?

Listen and inquire. Shut the door, unplug/turn off the phone, and listen. Oftentimes we want to do most of the talking, but solicit ideas from your subordinates on how they can improve themselves and their respective wards and sections, and you will have discovered the secret to effective counseling. Comprehensive initial counseling will set the stage for your expectations of your staff officers, and theirs of you.

Enable context. Give your subordinates several weeks to feel the ground under their feet so they have context before you sit down. Don't forget to provide your subordinates with your own OER support form, and that of your boss. Have them bring theirs to initial counseling. Hopefully, they will all be nested, but work together to produce a useful product.

Open the toolbox. The Army counseling statement (DA form 4856) is genius in its simplicity and effectiveness if used correctly. Ensure you utilize the Plan of Action and Leader Responsibilities blocks of the form. It doesn't have to be a pretty, typed document. Write on it, draw arrows, and connect the plan with your responsibility to them in supporting that plan. Much of it can be filled out during the session. You are, in essence, making a contract with that subordinate. Don't forget the 'Assessment' block at the bottom of every statement. Follow up with this at the next counseling session before breaking out a fresh form. Provide them a copy of the statement when complete with the session.

Expand your branch horizons. Whom have you counseled in the past? For most at the senior Company grade level, it's probably been a collection of officers/NCOs of similar branches (i.e. a tank Company Commander counsels tank and/or Infantry Platoon leaders, First Sergeant, etc.). As a Battalion Field Grade officer, it may be the first time in your career that you will be faced with the daunting task of counseling officers well outside of your traditional roles and experiences. This might include:

- Chaplains
- Adjutants (Adjutant General)
- S2s (Military Intelligence)

- Signal officers
- Chemos (there's more to being a Chemo than USR)
- Fire support officers (Field Artillery)
- Family Readiness advisor (civilian)
- Unit maintenance officer and tech (Ordnance)

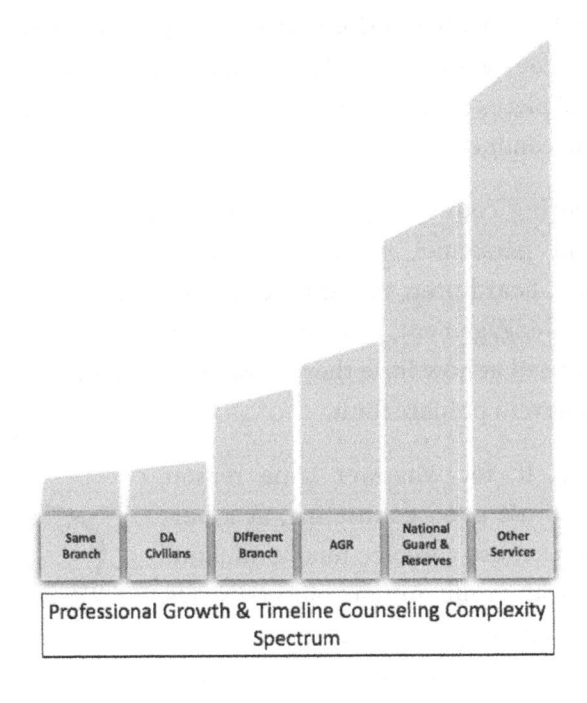

[Bar chart showing increasing complexity from left to right with categories: Same Branch, DA Civilians, Different Branch, AGR, National Guard & Reserves, Other Services]

Professional Growth & Timeline Counseling Complexity Spectrum

Scoring 'Distinguished' on Tank Table VIII means little to the AG S1 and even less to the unit chaplain. So, how do you sit down with officers of more 'eccentric' origins and conduct effective counseling? I would recommend doing the research ahead of time with Brigade Field Grades and senior warrants of their respective branches to determine how best to develop and mentor your subordinates. Try to get copies of Brigade staff counterpart OER support forms, and solicit input from them as to future positions, timelines, and professional development opportunities as well. If or when you find yourself working with National Guard or

Reserve Soldiers, DoD civilians, or members of other services, you may want to apply a similar research methodology to counsel effectively.

'Drop the Ramp' or 'Front Slope Counseling.' Don't pass up an opportunity to conduct verbal counseling or show an interest in your people anytime or anywhere. Pull a subordinate aside in the Joint Security Station (JSS) or Tactical Assembly Area (TAA), sit him/her down on a tank mine plow, or on the ramp of an armored personnel carrier, and 'AAR' his/her performance through a combination of direct and indirect counseling methods. If you are in a light unit and don't have any cool pieces of equipment to sit on, then I guess you have a good excuse not to conduct such counseling.

Go long. During counseling, provide subordinates with professional timeline and personnel file advice (future jobs, functional area opportunities, board prep, board procedures, records brief scrubs, DA Photo rules of engagement, promotion – command – schools, etc.). You may be surprised at how little they know about the future of their own careers and how to prepare for it.

Give closure. If, for whatever lame reason, you can't institute good counseling, at a bare minimum you must conduct final evaluation counseling with your junior officers/NCOs. Don't hide behind a PCS move and leave a Soldier in the dark by not telling them face–to–face how he/she performed in their service to the unit. As a rater or senior rater, you owe it to them, and to that leader's future subordinates.

Evaluation Counseling. I would submit that this counseling session can be one of the easiest, but only if the rater and senior rater have actively pursued a program of formal and informal counseling like the ones discussed above. But what is probably more typical, and surely more shameful, is one of the following examples: you receive an email notification to log on, review, and sign your evaluation before forwarding it to the S1 shop for processing; if you were luckier, perhaps, you were invited to attend that awkward OER counseling session where your senior rater asks you to sit down, then swivels the form around the desk for you to read in silence before asking if you have any questions. This is quickly followed by his or her indication of where you should sign (or that you can now log on and sign) accompanied by some token comments: "So, Don", "it's 'Dave', Sir", "Yea, Dave, uh... how's the wife?" "I'm not married, Sir." "Well, sure you're not, you lucky bastard! Gee... overall, great job, Dan, see you next year." Instead, make the time in these very important sessions to once again review important career milestones and timelines, as well as personal, professional, and section sustains and improves. Then look the rated leader in the eye and read your comments out loud, from top to bottom, as if you're debriefing the trooper's mother. Wrap up the session by soliciting questions or comments, and follow with a brief discussion of the way ahead over the next rating period if appropriate.

Plan Sleep

SLEEP IS A COMBAT MULTIPLIER. You may feel that as an Iron Major or senior staff leader, it's your job to outlast everyone or stay up through three shift-changes in the TOC. That may be the case, but only on the most extreme of occasions.

Sound sleep encourages sound decisions. Sleep plans are a matter of basic field craft and discipline. Some leaders require six hours, others eight (few need 12). Figure out how you function best, and try to stick to it. If you think staying up for 48 hours will make you more efficient and garner the respect of your subordinates, then you are probably oblivious to the poor decisions you made or the irascibility you demonstrated by acting like a cranky, soil-bottomed three-year old.

Sleep to think. Few things inhibit critical and creative thinking like a lack of slumber. It's extremely difficult for the sleep-deprived to think through the second and third order effects of an action, either on a wargame mapboard, or while managing the fight from the hatch of a tank. Even after a day or two of continuous operations, your ability to fashion inventive tactical solutions during a COA development session may often result in the lackluster 'frontal attack.' Whom would you rather have grab a scalpel to execute that vasectomy: the doc wrapping up a 30 hour shift, or the one who just got eight hours of sleep?

CTC ROE. Don't try to stay awake through an entire Combat Training Center (CTC) rotation; you're not going to stay up through a year's deployment, so why train that way? You will quickly hit the point of diminishing returns, and returns are what the CTCs are all about. Remember, it's a marathon, not a sprint—train as you fight.

There's no shame in naptime...mostly. While deployed, don't be afraid of power naps. It's hard to sleep the same hours every night in theater unless you are on shift, so make up for it when you can. Executing the Under-Desk Costanza nap in garrison is perhaps a bit more frowned upon, and rightfully so. But in a pinch, nothing beats the driver's seat of a tank.

Support the HHC Chain of Command

MANY OF YOU probably have had the sometimes unenviable job of the HHC/HHT Commander in the past, and can attest to the frustrations of dealing with an organization full of Captain-peers and Majors. For those of you that have been there, now's your chance to experience it from the other side of the fence. Remember the tough position the Company Commander and 1SG are in leading an organization with Field Grade officers and SGMs. Coach the HHC Commander on best practices for managing the challenges of this command. Ensuring your staff or section is meeting deadlines helps them immensely. A productive relationship with the HHC Commander and 1SG is essential for both

Battalion Field Grades and staff leaders, and as many of you know, there are plenty of friction points. Here are some examples:

HHC Formations. They can come in the middle of a staff IPR or at the 'end' of the work day, and few on staff would argue that they are not disruptive. But categorically exempting your officers from such gatherings often can potentially set them up for failure as well as establish a discouraging trend of 'RHIP' in the eyes of the NCOs and enlisted Soldiers in the unit. HHC formations will allow the chain of command to best distribute and collect information. Does that mean that every one of your officers is at every formation? Probably not, but talk to the HHC Commander and iron out an acceptable solution.

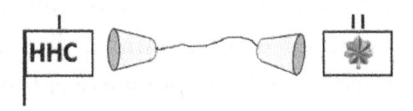

Company-specific events. The following are a few examples of events that require the staff's support and necessary crosstalk between you and the HHT Commander, and if they are not properly forecasted and coordinated with the efforts of your staff, they can become very disruptive to your priorities of work:

- Fitness tests
- Weapons qualifications
- Command maintenance & services
- Mandatory Quarterly Training (CO_2, POSH, etc)
- Soldier Readiness Processing (SRP)
- Weapons draw/cleaning
- Communications loading & change-over
- ISOPREP
- Holiday parties and other social events
- Family Readiness Group meetings, etc.

A good technique is to bring in the HHT Commander to one or two of your weekly staff huddles, preferably at the week's bookends, and share information. At that point, they can review Company-specific training events or other administrative tasks that your encumbered staff may need to attend or pursue over the near term.

Senior NCO communication. Managing the operations Sergeant Major and HHC First Sergeant relationship in regards to the S3 shop personnel may require the Field Grades to tactfully inject themselves to prevent a catastrophic breakdown that could become cancerous to the unit. Oftentimes the kinds of liaisons you have established with the HHC Commander are occurring in parallel between the OPS SGM and the HHC 1SG. This may be cause for confusion, thus having your OPS SGM at your staff huddles will be of some benefit as well.

Financial Liability Investigation (FLIPL)-instigators. Don't inhibit property accountability with loose practices. Your decisions as to the arbitrary movement of assets within your respective shops can set the conditions for lost or unaccounted for property (e.g., S3 yells, "my monitor doesn't work anymore, switch it with the Chemo's ASAP"). Arguably, the most commonly lost item is a laptop computer or monitor due to hasty repositioning within or between shops without setting the conditions for the hand-receipt paper trail to catch-up. Some events to pay close attention to:

- Turn-in of damaged items
- Re-imaging of staff lap-tops coming into and departing theater
- Chain of custody for staff/shop computers in TOCs or in transit (e.g., an S3 laptop set up in a TOC tent in Kuwait)
- Unplanned equipment swaps when items go down while in the field or conducting operations (generators, specialty tools, computers, weapons optics, etc.)

You can mitigate loss by rigidly supporting HHC inventory and accountability policies, and working all property book items or sub-component moves through the Section Sergeant who is responsible for the item. Ultimately, that's the guy who may end up paying for your haphazard property habits and mandates.

Predictability is the key to avoiding conflict & confusion. Create and maintain open lines of communication between you and the HHC Commander, and force him/her to put things on your calendar as far out as possible so you can synch your staff around them.

Master and Exploit Mission Command/ Command & Control Systems (and DTMS)

IF YOUR IDEA of executing mission command (formerly command & control) in today's fast-paced and ever-changing environments—COIN or otherwise—is through an acetate-covered map with push-pins, then you are probably doing your unit a disservice by not maximizing the potential of the Mission Command Systems at your disposal. Sure, analog back-ups are great, and even a necessity for most operations, but don't hamstring yourself and your unit by being intimidated by these systems. You don't need to be the primary Command Post of the Future (CPOF) operator in your TOC, but knowing the vast planning and situational awareness capabilities of the system will make you a more effective and efficient Field Grade. Ultimately, your expertise will show through on behalf of the unit through smooth BUBs/BUAs, effective parallel and collaborative planning, and a COP that allows the commander to make the best and most timely decisions possible.

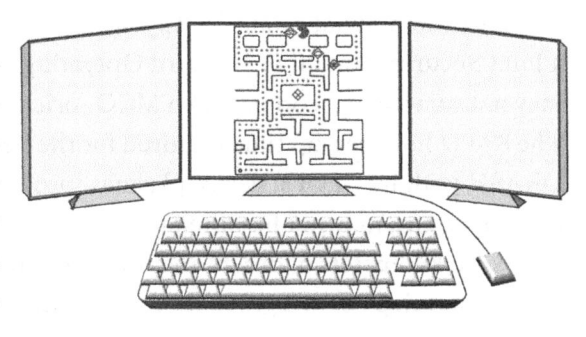

Get your hands (sort of) dirty. Don't neglect basic Force XXI Battle Command, Brigade & below (FBCB2)/Blue Force Tracker (BFT) proficiency; you may find this will be the best way to command and control an operation when the communications architecture won't support it in other ways.

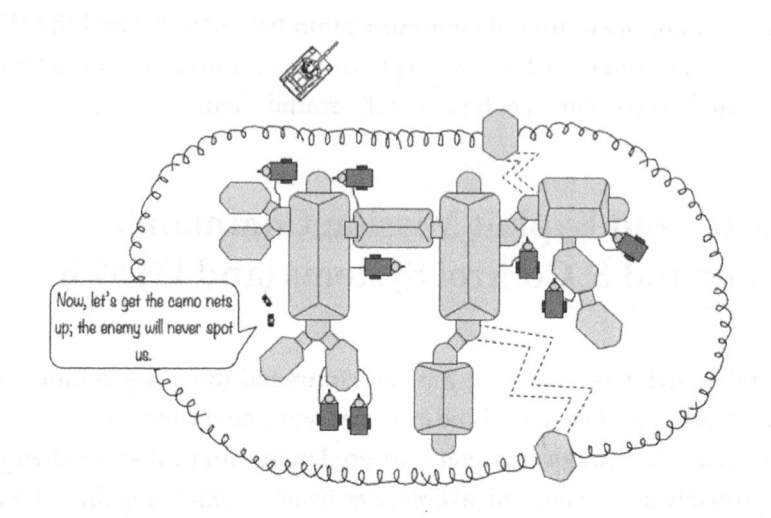

Now, let's get the camo nets up; the enemy will never spot us.

Strike a balance. If you refer to a trend of Forces Command (FORSCOM) Commander's Training and Leader Development guidance, you will see an increased focus on unit proficiency across the full spectrum of conflict. Battalion live-fires, combined-arms breaching and other perishable, Major Combat Operations (MCO)-related collective tasks are back on the Training Center and home station training menus. So, that hardstand 'steady state' TOC you worked in as a battle Captain on a Joint Security Station or Forward Operating Base for a year in theater may not survive first contact in an MCO-oriented exercise or operation. The 8 to 12 hour set-up time required for the behemoth, multi-bubbled DRASH with its CPOFs, desks, plasmas, projectors, break rooms, yoga chambers, etc., may not be suited for a Battalion attacking or sustaining across 150 kilometers of desert, regardless of branch or function. A TOC tends to develop its own gravitational pull after a while, attracting more tents, vehicles, chubby snacks, and yes, even enemy mortar rounds.

Think through, develop, and rehearse systems and SOPs ahead of time for a leaner MC structure that can teardown, jump, set-up, and protect itself rapidly to keep pace with a maneuvering unit. Do the same for the Rear CP and TAC, and don't forget a stout coffee pot in each.

Lost in Space: The Fickle Frontier

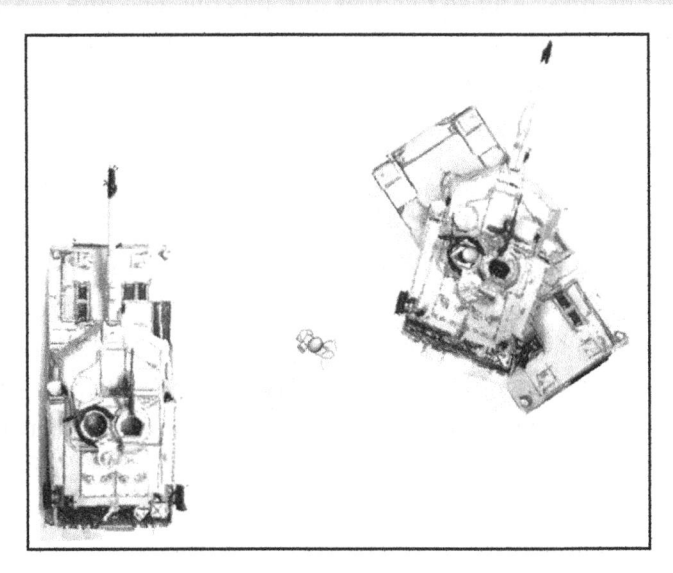

AS A COMPANY-TEAM and breach Company Commander during a deliberate night attack at the National Training Center, I recall an embarrassing episode when I was forced to jump tanks in the middle of enemy contact. As I scuttled on–foot (heaven forbid!) across the sand in MOPP II clutching my map board, overlays, and OPORD, I had no problem distinguishing the route to my wingman's tank with the help of my own tank's flashing whoopee light. Upon assuming this unfamiliar tank commander's hatch, I stared down blankly at the darkened display screen on the broken PLGR (GPS) strapped to the .50 caliber mount in front of me and I immediately began to regret my last hasty radio–issued instruction to the remnants of my Company. "Contact, damn it. Back up! Back up! Move 3000 meters east and consolidate vicinity Check Point 6; I'll link up shortly." In the pitch of night, lumbering around the desert in a solo tank sans GPS, it took me over 15 minutes to find my bugged-out Company despite transmitted grids and flashing headlights, all the while holding up the entire Battalion's attack.

The good news: there was a US News and World Report correspondent there to watch, record, and eventually publish the whole thing, much to my professional mortification (thankfully, I think the article has worked its way out of circulation by now, but I'm sure that, after this confession, some smartass peers of mine will dig it up once again). The point of the anecdote is evident: I relied heavily on my GPS, and when it wasn't there, I was rendered almost helpless (my suspect radio-issued retreat aside).

Now for a shameless yet wholly necessary and timely space plug. My recent transition into the Space Operations functional area after 20 years of tactical and operational Armor assignments has left me both enlightened (relative term, I assure you) and a bit apprehensive about our ability to fight as we're used to in future conflicts, both conventionally and unconventionally. Our attraction to the cushy Drash domes, myriad sophisticated computer systems, and Otis Spunkmeyer muffins continues to grow. And so too does our appetite for—and dependency upon—GPS-based and satellite-facilitated situational awareness, navigation, and mission command tools in our tactical operations centers, on our equipment, and in our formations.

Bowling Lane Mission Bumpers. Take a look around your TOC or Rear Command Post and make note of all the systems that rely on GPS or satellite communications to facilitate your mission command, regardless of your echelon. Now, let's turn out the lights. Park the Shadow, collapse your VSAT, unplug those Pluggers, put away your DAGRs, take off your (unauthorized?) Garmin watches, close down SharePoint, and douse those BFT and CPOF screens. Try to collect and distribute a situation report to higher now, blast the 2406, or better yet, initiate the Battalion's LD and orchestrate that nighttime complex obstacle breach. Scary situation? Maybe not in Call of Duty, but try it live and I'll bet there will be some buttered drawers and piddle-puddles around the Battle Captain stations, and much more from those prosecuting the fight from the hatch of a tank in the dead of night.

Let's face it; adversaries know how we like to fight. There's been a lot of note taking since Desert Storm, and we're sometimes not very subtle at disguising our glamorous toys, their capabilities, or their employment. Need substantiating evidence? Just pick up one of the many comprehensive white papers openly published by some of our 'international competitors.' That being the case, potential bad guys from around the globe, both sovereign entities and more obscure but wholly capable and sophisticated transnational groups have focused a lot of time, money, training, and technology into countering our reliance on the asymmetric bennies historically provided by our space-based assets. And, unfortunately for us, it's not too hard to do.

So what? Well, training in Mission-Oriented Protective Posture (MOPP) gear sucks. Practicing fighting in a chemical environment is hot, cumbersome, and wholly despised by all, except for the adversaries using it against us (and maybe the sadistic Chemo). Yet, historically, we assumed that every fight with our (former?) Bolshevik buddies would be fought so, and thus we rarely crossed the LD without the comforting heft of a quality charcoal suit. I would submit that we have to embrace the same mentality when it comes to fighting in a continuously or intermittently space-denied environment. It should be the norm rather than the exception that those sexy toys and capabilities should often be denied us in training, and always at the worst possible time. There are great Army space folks out there working to warn of potential degradation, and fighting hard to maintain our access to space, but regardless, train against it as if it were just another persistent form of enemy contact. To do anything else is merely to cheat yourself and your unit, and jump 'Hollywood style.' Routinely practice fighting without the space-crutch: develop analog back-up systems for information consolidation, production, and distribution; train the basics; dust off that compass thingy; and be prepared to party like it's 1999.

Fight the power. In addition to training without space, learn to incorporate mitigation countermeasures such as GPS-enhancing body and terrain masking techniques. Also be able to recognize symptoms of space degradation, disruption, or denial of your systems. At every echelon, ensure PACE plans (primary, alternate, contingency, emergency) are addressed for all types of communication within your command posts (FM, SATCOM, etc.). Seek out the latest Army Space Training Strategy for additional tips on how to mitigate adversarial counter-space efforts.

Conduct First-Class Ceremonies

I USED TO THINK ceremonies were a complete waste of everyone's time, so this one took me a while to internalize, much to my own detriment. But here's the bottom line: well-rehearsed, classy ceremonies are necessary traditions that reflect a well-disciplined and classy unit. Oh, and when a ceremony does goes south, expect that an embarrassed commander will promptly reorganize your priorities for you.

What's the payoff? Ceremonies such as Changes of Command, Changes/Assumption of Responsibility, Post-Deployment Battalion Awards Presentation, Reflagging, Deployment/Redeployment, Grog bowls, etc., can be resource and time intensive and can oftentimes distract from other missions or priorities. But have no doubt that they all contribute significantly to the identity, pride, and reputation of the unit, for better or worse. They also stand as means to render respect or honors for deserving Soldiers in many different capacities.

Plan well. Treat a ceremony like any other major training event or operation. Apply an 8-step Training Model approach and you will produce a first class event every time. Typically, a ceremony will (or should) involve the following: a deliberate OPORD published well outside of the near-term training schedule, IPRs, site reconnaissance, confirmation of resources, coordination with Protocol, invites & RSVPs, and lots of rehearsals, to name a few.

Unity of effort. Although your Operations Sergeant Major and/or the Command Sergeant Major may be the point guys or gals on this mission, especially with taskings, it will still require participation from and synchronization of the entire staff to pull together a good ceremony. Also, don't expect a random Captain in the S3 shop to construct something of this magnitude. Project officers are handy, but they can't go it alone.

Do the legwork. Protocol is important and integral to making people feel welcome. It's got nothing to do with ego and everything to do with etiquette and courtesy. If you have doubts, contact the Post Protocol office and get advice and answers. Don't set up your unit and your commander for potential embarrassment by not having your staff do the legwork ahead of time.

Well–rehearsed, classy ceremonies are necessary traditions that reflect a well–disciplined and classy unit.

Treat Attachments Like Invited Guests

MANY OF YOU have been there before, perhaps as a MITT chief in theater, a Sapper Platoon cut to the breach Company, an Air Defense section set in the Brigade Support Area, a Civil Affairs team supporting a maneuver Battalion, a Decon Platoon attached to a FLE, etc. Many were unlucky enough to be 'sliced' to a unit that treated your team as an afterthought and maybe even a nuisance. You were relegated to a peanut–gallery seat during orders development or targeting meetings, forced to dig up your own Class IX parts and supply items, considered second–class citizens at unit social events (if you were even invited), left out in the cold when it came to evaluation reports or awards. The list of affronts is vast. As one might expect, attachments' perception of a lack of support is soon reciprocated, thus creating a vicious cycle of poor motivation and degraded complementary or reinforcing capability—on the part of both you and your hosting unit. Interdict this cycle by aggressive integration of attachments into your unit, regardless of the duration and command/support relationship. Some recommended techniques might include the following:

The importance of Annex A. Understand command relationships and support relationships, and what's inherent in both. There are vast differences in expectations for TACON verses ATTACHED units, just as there are for DIRECT SUPPORT and GENERAL SUPPORT REINFORCING units. There are also plenty of issues that aren't clearly spelled out by the doctrinal definitions of both. Carefully analyze (gaining) host–unit obligations and attached unit (losing) higher HQ responsibilities. Determine, then plan and resource these requirements thoroughly. It will probably require a comprehensive staff mission analysis to tackle this from all aspects. Some common issues to work out:

- Billeting
- Food head–count

- Specialized maintenance support & Class IX flow
- Global Combat Support System (GCSS) operations
- Motor pool/tactical assembly area space
- Evaluations and awards
- Environmental Morale Leave (EML) tracking
- Mail distribution
- Army Direct Ordering (ADO) & distribution
- Field Ordering Officer (FOO) expenditures
- SOP & battle rhythm integration
- Reports and reporting procedures
- Tasking authorities
- Ammunition account management & resupply
- Isolated Personnel Reports (ISOPREP)
- Server/domain integration
- Medical Protection System (MEDPROS) & medical records
- Blue Force Tracker addresses
- Manifesting
- Investigation responsibilities
- Uniform Code of Military Justice (UCMJ)
- Commander's Critical Information Requirements (CCIR) notification & information flow
- Family Readiness Groups (FRGs)

Level the bubbles. Take a close look at the level of support your attached unit is getting from their own organic headquarters. If for some reason they are not receiving the same level of support as your Soldiers are, regardless of the specified relationship, your unit may have to make up the differences to ward off any perceived peer deprivation.

Make space at the adult table. A sliced Field Artillery Battery Commander & First Sergeant should get a spot at the table right next to the organic Infantry Company guys. Introduce and welcome new attachments to the audience, include them on meeting agendas, and give them time to brief.

Buy-in. Attachments are usually the smartest guys/girls in the room on their respective trades. Solicit input from attached leaders as to how best they can support the plan. This creates buy-in, inclusion, and loyalty, and optimizes their potential contributions to your unit's efforts.

> *Here's the bottom line: taking care of Soldiers is a universal responsibility, regardless of whether they're yours or not.*

Full-spectrum integration. Include attachments into unit social functions, both stateside and in-theater. This can range from unit BBQ's, organization days, award, patch, & spur ceremonies, unit birthdays, professional development sessions, staff rides, changes-of-command/responsibility, etc.

Here's the bottom line: taking care of Soldiers is a universal responsibility, regardless of whether they're yours or not. If you treat attachments like they are VIP special guests at Thanksgiving dinner, they will most likely bend over backwards to contribute to your unit's fight.

Master Army Property Accountability & Management

WHY SHOULD YOU CARE about property management in your unit when the only thing you and your commander need to be signed for these days is your TA-50 and laptop? One painful and costly acronym: FLIPL (Financial Liability Investigation for Property Loss). You, as the XO or senior staff leader, may still have a 'supervisory' responsibility in making sure the staff and unit follow sound property management procedures. If you are the XO, the boss (and Brigade) will be looking to you to manage FLIPLs within your respective organizations. With Company Commanders spending most of their commissioned time in and out of theater as of late, you may be surprised at how little some know about Command Supply Discipline. Close behind poor judgment and the loss of sensitive items, property mismanagement will get Company Commanders fired quickly, both CONUS and abroad.

> *Close behind poor judgment and the loss of sensitive items, property mismanagement will get Company Commanders fired quickly, both CONUS and abroad.*

Sound practices. Understand the following: shortage annexes, non-expendable shortage annexes, PBUSE/GCSS capabilities, supply item codes (expendable, non-expendable, durable), commander's write-off procedures, 10%/monthly inventories, and sensitive item inventories—in both deployed and garrison environments. Enforce a strict 'hands & eyes-on' approach for inventories. Temporary hand receipts or turn-in documents from the Defense Reutilization and Marketing Office

(DRMO) for Company night vision goggles getting purged does not equal 'item accounted for' on monthly sensitive item inventories. It's painful, but the inventorying officer needs to go to the workshop and lay hands on the items. If a deployed Company has a radio or computer turned in for repair, and it's still in theater, then ensure the inventory officer does everything he/she can do to get eyes on. If the inventory officer is someone other than the commander (regulation allows E7 and up), ensure that companies have certified and validated those inspectors as well.

No Free Chicken. Note that just because an item is expendable or durable doesn't mean Company Commanders can choose to disregard responsibility for the item. Many of these items are very expensive (impact wrench, ratchets, torque wrenches, etc.) and impending budget constraints will hold commanders more accountable in the future. The days of 'blank check' ordering for these category items are drawing/ have drawn to a close, even for deployed units. Influence commanders to hold hand receipt holders accountable and utilize statements of charge and write-offs when they are warranted.

When you deploy, don't let companies try to slicky-split property books. Don't let commanders try to cut a deal with a Rear-D Commander to store & inventory property book items. Connexes and headquarter facilities in the rear can get moved, may be turned in, or the Rear-D Commander will switch out while the unit is deployed. All of this may result in property book items getting misplaced and lost. It might require a comprehensive Rear-D connex inspection on your part to ensure no jack-assery has occurred prior to your deployment.

Book it. Your personnel will undoubtedly order and receive high-dollar items that won't come to the unit through the Property Book Office, especially in theater. Develop systems that ensure such items get on the books or they will walk. ACOGs, GPSs, hand-held radios, and off-the-shelf specialty tools are good examples of big money items that may 'exfiltrate' the unit upon redeployment if not properly accounted for.

Get your S4 shop in on training and unit QA/QC. Look for a great S4 NCOIC with plenty of years of Supply Sergeant time who can train and mentor Company supply guys and XO's, and conduct courtesy inspections of Company property management procedures on a monthly basis. This kind of scrutiny will minimize FLIPLs and keep you and your Battalion off of damning Brigade command & staff and deployment IPR poopy-stick-in-the-eye slides. Most Brigade and Division Commanders will insist that all FLIPLs are rectified prior to deployment & redeployment, so assign competent Investigating Officers (IOs) and keep the thumb screws down tight in order to get them done post-haste.

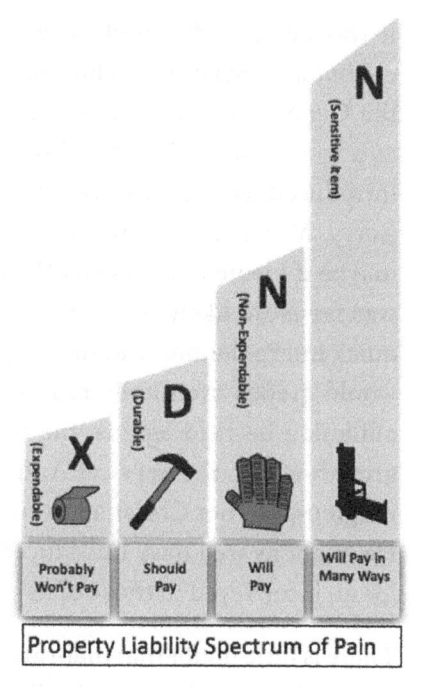

Battle track. Ensure you assign an officer (most likely your S4) to aggressively track FLIPL statuses and location. This includes further defining the nebulous location of 'up at Brigade.' Track whose office it's in and when it got there to include: 'with Investigating Officer,' 'at Company, Battalion command group, Legal, Brigade XO, Brigade Commander,' etc.

The Touchy Taboo of Religion in the Rating Scheme

As a Company commander, I recall an episode when one of my bosses, a deeply religious officer, was introducing members of his staff and command to a visiting General officer in our Battalion conference room.

CHAPTER II: THE ALGEBRA OF MAINTAINING GOOD HABITS **65**

It was obvious that both new each other well, apparently from their common association at church. I was quite surprised when he brought the General over to one of the Captains on staff and introduced him as a 'good Christian officer.' It got me thinking as to what I would be introduced as. Perhaps my single, partying, rare mass–attending ways (sorry, Mom) would garnish me the title of 'bad Christian officer,' or maybe if I were a tad luckier 'bad Christian, but good officer.' But what was far more likely to occur was just plain 'bad officer,' at least in my quick interpolation. And further, if that's what he really thought of me, would grades eventually reflect? Would my apparent tepid Christian affiliation be held against me in determining my potential for future growth within the Service? And, I silently conjectured, if I were stoutly religious, but not Christian, would that have been held against me too? He might as well have introduced that fellow as his 'Number One Guy' for all it conveyed to those of us within earshot.

There is little doubt that there are few more controversial and passionate topics like that of religion in the work place. It's damn difficult (no pun intended) to make a statement in this regard without disenfranchising some portion of the population, or without plunging into the rabbit hole on what does or does not violate the separation of church and state. So, that being said, I have no desire to interpret law or policy, or play devil's advocate, but I do think there is room for some mild constructive discussion. I will preface by reiterating my impressions as a junior officer in the anecdote above, thoughts that, upon further discovery, were reflected by a great many leaders within the organization across several echelons and ranks, much to the great detriment of our unit's cohesion and trust.

So, where is the line and when do you cross it? Is it a New Testament quote at the end of your email signature block, a picture of the Messiah on your desk, a Koran on your bookshelf? I would submit that none of these are inappropriate or even approach the line. Solders are often encouraged to attend and participate in unit prayer breakfasts, bible studies, and of course, military hosted services of all types. Plenty of great Soldiers embrace their religion, stand as stalwart examples of it—both in and out of uniform—and even serve as leaders within their respective congregations. Spiritual fitness is a vital component to the 'pentathlete warrior,' regardless of denomination or lack thereof.

The litmus test of inappropriateness is probably best taken from the perspective of the unassuming subordinate. At which point would one feel pressured, either directly or indirectly, to assume a role more in line with a supervisor's professed religion? At which point would a subordinate feel that their own faith or lack of faith served as a liability in the eyes of his or her supervisor? It has been my experience that it takes only a perception of such partiality that can be detrimental to the staff or unit, much less an actual or tangible deed or trend. In the anecdote above, I genuinely felt disenfranchised as a result of what I heard, and, after further discussion, it became readily apparent that a great many more of us felt the same way. Be aware of how your actions or lack thereof might be perceived by all walks of life within your ranks, regardless of religion, ethnicity, gender, or orientation.

Chapter III

The Trigonometry of Staff Management

THIS CHAPTER INCREASES the task complexity and challenge yet again by examining important aspects of organizational leadership that can be decisive in molding excellent staffs. Over the next couple of chapters, you will perhaps start to detect an underlying theme, one in which people and our ability to build and maintain relationships with them begin to differentiate between those who can merely get the job done satisfactorily, and those who truly evolve and transform both their units and themselves. It's time to separate the wheat from the chaff.

RIP a Good One

FEW PEOPLE LIKE conducting an individual or counterpart Relief-in-Place (RIP), regardless of whether you are coming or going. For the outgoing person, the new guy is a persistent, nervous close-talker that asks too many questions, disrupts your daily priorities of work, and takes up precious room in the office while constantly violating your personal space bubble. For the incoming person, the old guy is a cocky, self-absorbed fellow who won't tell you enough,cares little about your great ideas, doesn't share his computer, and can't seem to get out of your new chair quickly enough. As an arriving staff leader, there will be fewer more important days over the course of your tour than these that will best enable you to quickly perform well, and without having to reinvent the wheel or live-out the sins of those past generations. Still, the onus of a successful transition—and thus, successful continuity of operations—mostly rests on the shoulders of the old guy. So, let's expend some text discussing the keys to success for any good RIP, from both vantage points.

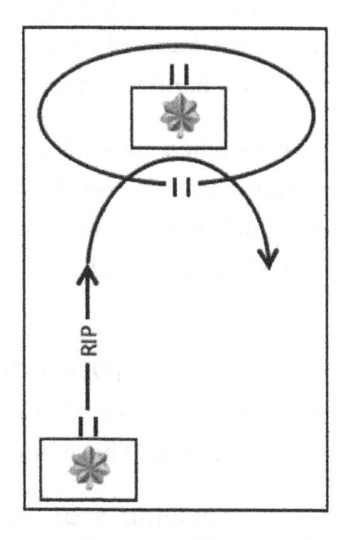

There are several kinds of RIPs, and some enable better transition than others. It may be as easy as taking over for a leader who is merely assuming another role on that same staff (e.g., an incumbent vacates the S3 job to assume an XO role in the same Battalion). In this instance, the predecessor is always on hand—for better or worse. In other situations, the predecessor may be moving up to a job on a higher echelon staff (e.g., a Battalion S3 moves up to assume the Brigade S3 role). In cases such as these, there frequently is an enormous amount of pressure on the incumbent to quickly wrap up any RIP, and get up to the higher HQ ASAP to begin the presumably more challenging role. Often enough, the incumbent may be PCSing to another installation or has already, which can complicate continuity of operations that much more. In some cases, the inbound trooper is serving as an individual augmentee, deploying forward to replace someone in theater who might or might not be on hand when he or she gets there.

As mentioned above, there are often many instances when the inbound personnel don't even have the benefit of seeing the outgoing ones at all, much to everyone's detriment. But, if you and your unit are lucky enough to have both on hand for any length of time, ensure you make the most of it.

Damn Old Guy (DOG) Rules of Engagement. Let's empathize a bit with the erudite DOG. You are probably coping with several priorities simultaneously. These most likely include:

- Managing your current battle rhythm to include projects, meetings, boss due-outs, etc.
- Preparing yourself and/or your family for the iminent PCS move
- Laying the ground work for your future position
- Perhaps even 'dual-hatting' between your current and future positions
- The increasingly painful process of clearing your installation

In addition, you must now receive, sponsor, train, and successfully establish some stranger into your soon-to-be former job. You're also forced to execute this tough task under the increasingly insecure guns of your boss: "that new guy better know how to [insert task here], or I will find you—it's a small Army." Such positive motivation makes you care for the bug-eyed FNG that much more. As you wrestle with these priorities, keep in mind that you have probably just spent the last couple of years of your life trying like hell every day to make your unit better. Don't tarnish those efforts now because you failed to adequately prepare the new guy to continue your legacy of excellence. With that, consider the following in preparing the FNG and your unit for transition:

Prep the objective. Don't expect the incoming guy to retool your Annual Training Guidance before he arrives at the unit, but there are probably some key documents or slide decks you can send ahead of time to better prepare him for the office upon which he is about to enter. Some apropos light reading will provide indispensable background knowledge and context that will allow both of you to advance the speed and quality of your RIP. Consider sending the following:

- RIP schedule
- Battle rhythms (all applicable echelons)
- Outlook calendar invite—if possible (or screen shots)
- Short and long range training calendars
- Unit organization wire diagrams
- SOPs
- Battalion and Brigade training guidance
- Command philosophies (Battalion, Brigade, and Division)

Digital and physical access. Within the numerous constraints imposed upon you by totalitarian IT regimes and security office beefcakes at all echelons, attempt to establish required email accounts, parking passes, and access badge applications ahead of time to facilitate rapid access to network, SharePoint/share drives, and facilities.

Plan the RIP. This task should not be a 'react to contact' or just a 'follow me' operation. Plan this event as if you were developing an orientation and integration for the new Brigade Commander. Identify objectives and create a detailed calendar and daily schedule that includes such events as:

- Base geographic orientation
- Unit footprint & facilities tours
- Your own section/staff personnel orientation
- Higher staff and adjacent unit peer meets (don't wait for a meeting, execute on their turf)
- Subordinate unit orientations
- Base supporting personnel meet & greet

Subordinate staff section briefs. These will be important meetings for the new guy's soon-to-be minions. Give general guidance as to what should be presented to your successor, and proof their products ahead of time. But grant them latitude—and maybe even space—in their presentation and delivery. Your presence at these briefs may stifle questions from the new guy, or quell candid statements from your subordinates. Plus, you could probably use the time to check email, work issues, or cuddle with the boss one last time. Have sections preface their briefs with a quick tour of, and orientation to their respective shops. Guidance for the composition of such briefs might include:

- Section/staff/unit MTOE or TDA and C2 wire diagrams
- Personnel lay-down (authorized/assigned, time in position, projected losses, etc.)
- Key equipment/vehicles and slant/status
- Capabilities
- Daily/weekly/monthly battle rhythm
- Reccurring due-ins and outs
- Ongoing projects and priorities of work
- Current/projected issues and challenges (important)

Hip-pocket training. Have an on-hand stack of required readings and briefings for the FNG's reference when he or she gets there. These may prove helpful when you need some maneuver space to fight the current fight. This collection should probably include:

- Applicable doctrinal pubs (narrow it down from simply, "read ADRP 3.0, you'll love it; now go play")
- Your unit's and next higher's SOPs/TACSOPs
- Commander's Critical Information Requirements (CCIRs) (multiple echelons)
- Commander's policy letters and philosophies at every echelon
- Unit-specific online classes or training requirements (IA, DSCA Phase I, etc.)
- Training calendars and schedules
- Recent meeting and IPR slide decks (last week's Command & Staff, Training Meeting, Unit Status Report (USR), Combat Training Center (CTC) deployment prep, and AARs)
- Lessons learned archives
- Command Inspection Program (CIP/OIP) out-briefs, etc.

If required, respect product classifications and track documents while ensuring the (properly vetted) FNG knows how to access and store secret material.

Air out those skeletons and skid marked tightie-whities. You've undoubtedly worked hard over your tour, and are probably damn proud of your clever systems and intuitive products. Spend necessary time showing off such gems, but be forthright with what tasks went unloved, and what traps or ambushes might be lying in wait for the new guy as a result of some neglected priorities of work. This might even include a list of tarnished relationships around the post that could benefit from a 'fresh face' in the position.

Establish working space for the incoming guy to operate from. A workplace or isolated cubicle or desk with an office starter kit, NIPR/SIPR computer, and printer access will allow the new guy to get and stay organized, and will grant you some much needed room to breathe. This may even require you to bounce some disgruntled officer from his or her desk for a week or two.

"Good luck with this dork, fellas." Don't slander the new guy to your old crew, as it will take exponentially longer for him/her to dig themselves out of the hole you put them into, whether justified or not.

Just the facts, ma'am. Determine just what it is you want to tell—or rather what's appropriate to tell—the new guy about your personnel. As part of that discussion, don't forget to review rating profiles, support forms (yours, your boss, and senior rater), and personnel files. Refrain from rendering categorical judgment on your soon-to-be former subordinates. "Captain Dunphy is a mouth breathing window licker" is probably not the constructive and insightful analysis the new guy was hoping for. Perhaps a better approach would be, "Dunphy has been late to work three times in the past month, his performance is substandard a majority of the time, and he chronically misuses semi-colons; here's his counseling packet." Upon your departure, the FNG gets to determine if he/she wants to have the 'clean slate/reset button' discussion with that suspect Dunphy guy, based on his or her own leadership philosophy and, potentially, input from the commander.

> *"Captain Dunphy is a mouth breathing window licker" is probably not the constructive and insightful analysis the new guy was hoping for.*

Second stringer. As the DOG, it's not out of the question to assign a staff officer to escort your replacement or take care of some aspects of admin to include in–processing, facilities orientation, etc. This frees you up to focus on some of those other priorities, and provides a chance for the new guy to meet or get to know some of his or her future subordinates.

Ride transition. Don't drop that ninja smoke bomb too soon when the switch from right seat to left seat occurs, but don't be too obtrusive either. Remain subtle with your guiding hand, and restrain yourself if the FNG wants or needs to flounder about on his or her own. Sometimes the newbie will be more comfortable with your absence, especially with the added security of some easy reach–back capability when those inevitable tough questions requiring some tribal wisdom and context arise.

Reach back. When you do finally bail out, make yourself available to the FNG whether on leave or at your next duty station. On occasion, be proactive and reach back through a quick email or phone call to ask if there have been any issues or unresolved questions that arose after your departure. It's a classy move and shows you care about their success and that of the unit. After a month the FNG has probably got it, but be prepared for that panicked call prior to the IG inspection.

Other DOG Bones:

- Provide a glossary of unique acronyms, organizational charts, and wire diagrams with names and positions ahead of time to better facilitate retention for the overwhelmed newbie.

- Arrange meet and greets ahead of time with busy folks so you don't drive across post to meet the Director of Training, only to find her out on leave.

- Don't forget that he or she is 'drinking from the fire hose' with all that you present him or her, so build in time and space for some internalization, note–taking, and reflection between events.

- Brief the boss on the plan in order to secure buy-in and the maneuver space to execute.

- Provide a copy of your OER support form that includes an updated job description. Throw your rater/senior rater forms in there as well.

Friendly New Guy (FNG) Rules of Engagement. One of the many beauties about our culture is the fresh set of objective, critical eyes each position gets every 12–24 months. The drawback, of course, is the lack of continuity and historical context instigated by a never-ending revolving door of FNGs. This lack of context often contributes to the reliving of past sins in a vicious cycle that can result in the tragic loss of hard-won wisdom and lessons learned by previous generations. As the new guy, your primary mission is to inherit the torch and tribal knowledge of the DOG, to ensure that you and the unit don't fall victim to this frequent trend. So, from the FNG's vantage point, let's discuss some keys to your success...

Don't render judgments. When provided with counterpart, staff, or sub-section information briefs, ask intelligent questions and feel free to drill down into their respective areas to understand the situation. Shy away, however, from making statements or rendering judgments on their products, priorities of work, or chaotic systems. Refrain from statements such as "I don't care for that much; I'll be changing that shortly." Take some notes and address it when you get put in charge (I would recommend soliciting some buy-in first, from both your minions and your boss).

Keep a low profile around your future wares. Feel free to engage in small talk and light conversation, but don't get too chummy with members of a staff or section that don't answer to you yet. Don't take up their time or interfere with their priorities of work or directions from the DOG either. Do visible PT, but don't be in rush to thrust yourself upon the group.

Grace us with your wisdom…but not quite yet. If you are sitting in a meeting with the outgoing guy and his/her subordinates, peers, or bosses, don't be in a rush to caveat or contribute without being solicited to do so. Even then, tread lightly. Smile and nod at all the right places. You'll get your chance soon enough, Neidermeyer.

Mind the trigger. Just like an actual unit relief in place, there is a set of conditions that triggers the positive hand–over of the fight from one leader to the next. The same is in effect for you and your predecessor. Don't give any direction or philosophy to your new staff until that trigger has been met, and you assume the position. They don't answer to you until that day, so defer unto your predecessor, and don't violate that often–neglected Principle of War: 'Unity of Command.'

You're it. Once you do formally take over, you are in charge, so be gracious and act like it (without coming off like a cocky know–it–all). However, don't be afraid to solicit feedback from the outgoing guy when you get a chance. In addition, the staff has to understand now that there's no more taking a slide or memo to the old guy for a final look or a bit more guidance. At least, not without asking you first. You have the con.

Don't bad mouth the outgoing guy or regime. That's a fellow officer/NCO for one thing, and there is probably (hopefully) a legacy of some loyalty, respect, kind feelings, and possibly a sense of loss within the organization felt upon his/her departure. It is a distinctly classless and unprofessional thing to do, and you may well instigate some sideways glances, furled brows, or even thinly veiled hostility from your subordinates by doing so.

Assess, assess, assess. Don't be too quick to affect a tectonic paradigm shift within the staff, section, or unit. Gain context over a few weeks and conduct formal or informal sensing sessions to better understand what might or should change and what needs to or can stay the same. Be savvy in constructing sensing session groups. Try to organize these by rank/grade, and peel off supervisors from subordinates to facilitate free and unencumbered discussion.

Gain context. Understand first, solicit recommendations second, and then vigorously execute rapid change (within your span of control, and with the blessing and guidance of the boss). Then reassess. Don't wait too long to enact perceived change for the better; before you know it, you'll be off to the next job.

DOG-free RIP. What do you do if no one is there to replace (i.e., the outgoing guy or girl has already shoved off)?

- Take charge now!

- No free chicken. The often-forgiving weeks you might have been granted as a Second Lieutenant to shyly probe around your new career field, determining what works and what doesn't is a luxury your unit can't afford from a new Field Grade officer or senior staff leader. You must perform and produce immediately.

- Rapid, 360-degree situational awareness and understanding is even more essential to your success and productivity. Modify the OPTEMPO to facilitate your situational understanding. This may mean daily staff/section stand-ups, huddles, more comprehensive rehearsals & pre-briefs, etc. You can slow down/ ease off the throttle later as you get more confident in your surroundings and your personnel.

- Lean heavily on your peers within the unit and those in similar positions in adjacent units to facilitate your understanding quicker, but do so when convenient to their busy schedules.

- Sensing sessions by staff section must be conducted with a sense of urgency. Also, inquire from adjacent units, higher staff, and even your boss as to what, from their respective vantage points, needs to be done better within your shop (without belittling or badmouthing the departed).

- Don't fall into the trap of over-relying on one of your subordinates to 'carry' you. If you have to, spread the wealth of making up for your ignorance equally across your new wares. It's not a bad thing to tell your folks that you might need to lean on them until you get your feet up under you. They still, however, should be able to rely upon you for lucid, informed guidance and sound judgment.

- Keep things in perspective, realize where you are and what you're doing, and don't forget to enjoy yourself. Yeah, it will suck at times, but at the end your tour, you will dread its conclusion and bristle at the thought of handing over those precious reigns to anyone else.

Manage Your Staff

LEADING ≠ MANAGING. There are overlaps of course, but understand the finer differences. Figure out how to assign tasks, give guidance, establish suspenses, follow up, and control quality. It's easy to hand out tasks, but it's infinitely harder to keep track of them all and follow up in a predictable, timely fashion.

Develop a system of task management. It might simply be dog-eared pages of your worn out Steno notebook with checked boxes & bullets or a digital system like Microsoft Outlook. I recommend a combination of both. Learn about and utilize the task management tools that Outlook has; it will save time, help you track, and prevent 'dropped balls.'

Spread the wealth. Don't be afraid to give staff leads tasks outside of their doctrinal range fans, especially if the primary officer is incapable or overly tasked already. Example: when the S1 is up to his armpits in developing the manifesting for a deployment, hand over responsibility for managing the staff duty officer (SDO) to the S6 signal officer (SIGO), who may be less decisively engaged at the time.

Manage *through* your minions. Don't get comfortable bypassing your subordinates to go to their subordinates. Few practices emasculate your junior leaders like making this a chronic habit. At times you may find yourself in need of a quick answer from a staff section, and, in the absence of the staff lead, a drive-by on one of their subordinates for a quick RFI is certainly acceptable. However, continuously doing so only makes your immediate subordinates feel irrelevant, especially when you start to routinely issue tasks directly to their underlings. If there's a lack of trust or confidence between the two of you, then conduct the required counseling, coaching, or re-training to rectify your subordinate. If that fails, then pursue repositioning, removing, or replacing the staff lead. When conveying tasks, requesting information, or assigning action officers, do so through the respective staff section leader. Empower your leaders by letting them do the analysis to determine how best to answer the mail and comply with your clearly issued guidance.

Dig deeper. For a quick and effective study on essential organizational management skills, I recommend reading _The One Minute Manager_ by Kenneth H. Blanchard, Ph.D. and Spencer Johnson, M.D., and _The One Minute Manager Meets the Monkey_ by Kenneth H. Blanchard. Put both on your staff's mandatory reading list as well, and discuss over an O/NCOPD.

Empower your leaders by letting them do the analysis to determine how best to answer the mail and comply with your clearly issued guidance.

Lead Your Staff

"NO KIDDING," YOU SAY. I'm not going to lecture you on leadership; you know it already. But don't be surprised at how you might compromise your own leadership philosophies when you assume the role of a staff section leader or "Iron Major," and get caught up in the race trying to produce slides, meet deadlines, and take care of the unit and boss. That philosophy you spent so many years perfecting may devolve into something as simple and uninspiring as "row well and live, Soldier." Remember that, in addition to managing your staff, you still have to "lead" your staff. Many a good junior officer has decided to bail on the Army because of a bad experience on a staff, most of which were instigated by a manager who didn't care enough to lead them.

> *Many a good junior officer has decided to bail on the Army because of a bad experience on a staff, most of which were instigated by a manager who didn't care enough to lead them.*

T.I.P.S. When I was a Platoon leader, then MG Swartz, the 4th Infantry Division commanding General, created and posted 'T.I.P.S.' signs throughout the post. I still find it a useful acronym today: Talk to (or rather, with) Soldiers, keep Soldiers Informed, be Predictable, and be Sensitive to Soldiers' needs. It may sound a little cheesy, but it's spot on, especially for a busy staff leader. Don't fail to remember that despite getting paid the 'big bucks' (a misleading term as you well know), junior officers are Soldiers too. Recognize their efforts in front of the boss, put them in for awards, counsel them, and care about their well-being and that of their families. Do the same for your Soldiers, NCOs, and civilians. Staff work can often be a thankless endeavor for many; don't let it be for your troopers.

Develop a Staff Leader Philosophy

HANDING OVER THE GUIDON before assuming your role as a staff leader doesn't—or rather shouldn't—exempt you from producing and abiding by a continued leadership philosophy that will provide some parameters, guidance, and predictability for your staff or staff section. Issue this guidance as you would in front of your Platoon or Company, with all hands present. Then be prepared to expound upon your tenets or answer questions when complete. I pilfered, created, tweaked, and (sometimes even successfully) employed the following excerpts as a staff leader in the past. Perhaps they can be of assistance in your efforts to do the same. This is a bit more of an intellectual approach than a simple list of pet peeves (although there is benefit in those as well). I tried to take my philosophy to levels beyond the mere "I hate phone-Colonels and paperclips" kinds of comments. I would also submit that a staff leader philosophy is justified in exceeding the 'one page rule' of a typical leadership philosophy. I found it helpful to organize my tenets into three broad categories:

- About me—necessary adages to better enable my subordinates to know the terrain/boss (enemy?)

- What I expect of you—the staff's daily obligations and rules of engagement

- What you can expect of me—my obligations to the unit and each member of the staff

This guy.

- My misgivings (among others): I'm absent minded, often disorganized, and tend to fly by the seat of my pants. I'm impatient. I'm a poor proofreader, and I don't focus enough on the details. I enjoy the 'close fight' and firefighting, and tend to neglect the 'deep fight'—all attributes that, in your absence, might contribute to our section or staff's failure.

- This may seem to contradict the last item, but I appreciate predictability, especially in my daily & weekly battle rhythm.

- If I look busy or harried, I probably am. Wait for a break-in-contact before approaching me with an issue, or you're likely to get some poor or vague guidance from me. This does not apply to Commander's Critical Information Requirements (CCIRs).

- Although guilty of it myself, I confess that I often have little time or patience for loquaciousness (this document aside). Be clear and concise in your questions, answers, emails, briefs, and memos.

- I believe that one's experience can be both an asset and a liability; be cognizant of how you leverage yours.

- I appreciate critical and creative thinkers. Avoid creativity pitfalls such as "this is the way we've always done it", "this is what the book says", group think, etc.

- I am an optimizer by nature. I want to make the most of every person, hour, opportunity, and dollar. I tend to question the norm, the status quo, and "the way it's always been done." I constantly work to seek better ways (not necessarily easier ways) to task/mission accomplishment.

- Staffs exist to support commanders at all echelons within the organization.

- Visual products sell a concept better than any cleverly composed thesis or monologue.

- I believe in well-prepared, rehearsed, and professionally conducted meetings & briefings. Meetings are created to scratch an itch somewhere; there must always be a stated purpose (information, decision, other?) and agenda; no decision briefs or 'gotcha' slide shows without a read-ahead.

- I believe that none of us is smarter than all of us; collaborative and parallel planning should be the norm, not the exception; relationships matter.

- I work hard and play hard, but I'll keep it professional in or out of uniform, with or without a beer in my hand.

- I am a traditionalist for military customs, courtesies and etiquette. When a person chooses to ignore these, I find it disrespectful to both the Service and the object of his/her ill-advised conduct.

- Keep things in perspective. Few crappy days here in garrison can rival a bad day in theater. No one's shooting at you or planting IEDs on your route home, and you have running water and porcelain within easy grasp. So, when the chips are down, step back briefly and grasp the greater goodness that surrounds you. Then get cracking.

What I expect of you.

- Work to make your respective areas better than they were when you got here. You should strive to get to a point where your successor has nothing to do but execute and monitor your refined systems and SOPs.

- To achieve the goal above, start dedicating effort now to your section's and your own position's continuity. Organized, intuitive digital products should be the standard you strive for to best set your followers up for success.

- Physical training is part of our Warrior Ethos and our way of life. If it's not part of yours, make it so…NOW!

- I expect suspenses and deadlines to be met. If, pending your own mission analysis, you determine that you are unable to achieve a milestone, propose a realistic 'way ahead.' But do so in a timely enough fashion.

- I don't want to be the one who has to follow-up on a task handed to you; don't make me be the nag. When I hand you the monkey, I expect you to take ownership, coddle that hairy critter, and keep me informed of its status.

- If you do not understand my intent, guidance, or direction, seek clarification immediately. I would rather have you question my vague and ambiguous statements or illegible dry-erase board doodles at the point of issue than have you flounder about in uncertainty and produce an inadequate product once we're at the LD.

- Account for, take care of, and recognize your troopers, regardless of rank, responsibilities, or position.

- Find sound footing somewhere between—and far from—being a 'yes man/woman' on one extreme and a 'fight the power' guy/gal on the other. When solicited, be candid; state your opinions honestly and tactfully. Sure, let's debate, but once a decision is made, I expect you to own the task, and execute it with vigor.

- When and how long you and your people work is somewhat up to you (within logical range-fans). I don't render judgment on an individual's ability or competence based on the number of hours they toil in a cubicle, but rather on how efficient and effective he/she is while at work. If, however, you find yourself not decisively engaged, march to the sound of the guns.

- Refrain from saying 'no' to subordinate units or adjacent staff sections. Do the analyses first, determine the risks and trade-offs, and strive to contribute to the betterment of that unit in any way that you can.

- Good staff officers/NCOs provide recommended solutions to problems and challenges (almost) every time. However, if you are drawing a blank, don't hold back on informing me of an issue simply because you can't envision a viable fix.

What you can expect of me.

- Respect
- Honesty
- Candidness (occasionally tactless)
- Frequently short attention span
- I will attempt to lead and manage through you, and by empowering you whenever I can
- Along with assigning tasks, I will also strive to provide you with:
    > Suspenses
    > Intent
    > Guidance
    > In-Progress Review opportunities
    > A final QA/QC
- I will strive to provide you with personal feedback through:
    > Formal and informal counseling
    > Performance Evaluations
- I will strive to provide professional development through:
    > Formal and informal counseling
    > OPDs
    > Occasional homework
    > Involuntary mentorship
- When I fail to accomplish these tasks to an acceptable standard, please let me know.

Staff Choagie Professional Development (SCPD)

JUST LIKE YOUR COUNSELING EFFORTS, professional development programs somehow work their way down to the bottom of the to-do list for those leading a staff. Interestingly enough, as with counseling, there are probably few better uses of your time to invest in the possibilities of exponential gains for your unit, your subordinates, and your subordinates' future units.

> *The only constraints on a good staff program are those created by your own lack of imagination, motivation, or mismanaged priorities.*

Often enough, the only formal professional development your staff officers and NCOs receive is that dictated at the Battalion program level—one championed by the Battalion Commander or Command Sergeant Major. This curriculum is often high quality, but perhaps too infrequent and general for your staff officers and NCOs to benefit from as a stand-alone program. The HHC Commander is also probably not in the best position to design or execute a staff-centric professional development program. But there's no doubt that your proximity to your wares and the ability to manipulate the training calendar provide you with ample means (and the onus) to construct one of your own. The only constraints on a good staff program are those created by your own lack of imagination, motivation, or mismanaged priorities.

Successful programs can often be executed with little overhead, with the exception of our most precious commodity: time. So, based on our reluctance to dedicate this resource and our traditionally languid approach to staff professional development, perhaps there is warranted room for some additional discussion on the matter.

Start point. There are many different types and methods of professional development programs, and they can range in complexity and resources like that of any other training regimen. Just like other training initiatives, the first step is to collectively assess your junior leaders to determine a suitable professional development curriculum, and then deconflict topics by nesting your program's efforts with that of other echelons, both above and below. This will ensure that your troopers don't suffer through two or three identical OPDs on 'filling out an NCOER.'

A range of options. What areas could your staff benefit best from? Based on their experience and past trend performance, these might include such unglamorous but often-necessary fundamentals as:

- Memo construction
- Staffing & routing documents
- Microsoft Office Suite skills improvement (PowerPoint, Excel, and Word)
- Conducting a 15–6 or financial liability investigation (FLIPL)

More sophisticated generic subjects might include:

- Preparing for the promotion/command/school/SER boards
- Conduct of a DA-level board
- Utilizing evaluation support forms
- The Army profession and living the Army Ethic
- Completing evaluations (officer/NCOs/other services, when warranted)
- Composing an executive summary (EXSUM)
- How to counsel subordinates (high payoff)
- How to develop an effective command or leader philosophy
- What it means to 'take care of Soldiers'

Increase the complexity a bit more by conducting and briefing analysis of historical battles or campaigns utilizing the Principles of War, Tenets of Army Operations, Characteristics of Offensive Operations, etc. Task subordinates or small groups to develop and brief articles for submission to military journals (this can also serve as a great writing assessment tool for you). And every good professional development program always has room for collaborative discussion on intelligence preparation of the battlefield (IPB) and the Seven Steps of Engagement Area Development (good for all branches!).

Prep = (payoff)^2. Regardless of what topics you chose to embark upon, prior preparation from both the facilitator and the audience will maximize payoffs for such events. Assign homework well in advance that might include doctrinal or historical research, or consumption of and reflection upon a recent professional journal, newspaper article, or book. If no context or general knowledge is established for the audience beforehand, your attempted interactive discussion will quickly devolve into the dreaded and uninspiring, head-bobbing lecture.

Customize your audience. By nature, battlestaffs present a challenge in determining what areas to focus on. The eclectic origins (branches) of your officers and NCOs might render that 'How to Fight with Tanks in Urban Environments' discussion somewhat benign to a large part of your audience. Consider tailoring a 'program of instruction' that might have select audiences for some sessions (officers, NCOs, soon-to-take Company command troopers, etc.), and broader topics that can appeal to all of the masses for other sessions (admin skills, battle analysis, article publications, etc.).

Tactical decision games (TDGs). Successful professional development programs at the tactical level might include tasks ranging from generic topics like those discussed earlier to more specific, branch-oriented subjects like movement or maneuver-focused micro-armor vignettes or tactical decision games. Tactical Decision Games are probably unrivaled in their prep-to-payoff ratio, especially for smaller groups of officers or

NCOs. You may only need about 20 minutes of prep time to conjure up a tactical dilemma or scenario, and sketch it out on paper or dry-erase board. Provide your minions with 30 minutes to derive a quick solution or FRAGO, then pick some wall-hugger to pitch his solution to the group as if he were fighting from the hatch or wood line. AAR and discuss, then move on to another victim. This low-overhead development technique can be used for audiences of almost every branch, and the resulting discussions can best be described as priceless. What better way is there to harmlessly train and discuss tactics, decision-making,

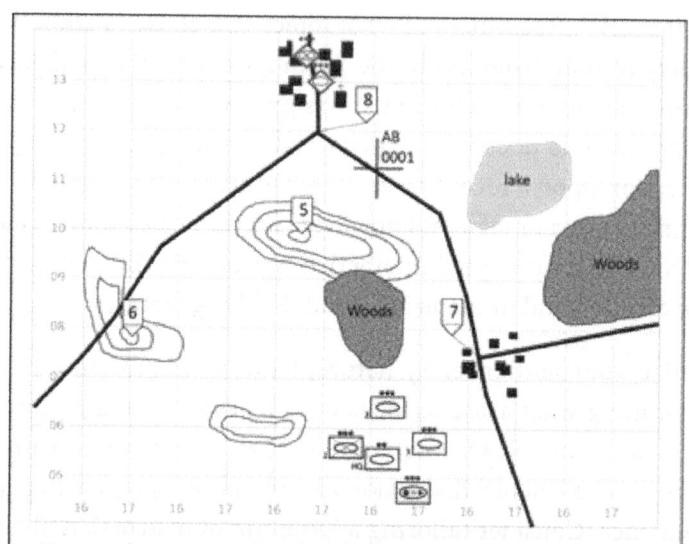

You are the commander of Alpha Team (2 Armor, 1 Infantry, 1 Mortar PLT). You are the advance guard company for your battalion's movement to contact. Your mission is to attack to destroy the enemy's lead company VIC CP8 in order to provide maneuver space for C Company, the battalion's main effort. Your unit is about 3-4 K's north of the battalion main body when you receive a spot report that the enemy MIC(+) has advanced further south than expected, and a T-80 and BMP platoon are rapidly approaching CP8. Keeping in mind your company mission statement, you have 15 minutes to develop a FRAGO and any supporting graphic control measures. Upon completion, be prepared to issue your FRAGO to the company over FM.

Tactical Decision Game Example

90 THE IRON MAJOR SURVIVAL GUIDE

creative thinking, and critical analysis where the only victim might be someone's temporarily bruised ego while you all collectively pick apart their plan?

Lead and learn through presence. As a staff leader, don't simply arrive at your staff's professional development events to deliver a couple of inspiring introductory comments, then abruptly about face and bug out. By doing so you merely convey its overall unimportance, and with your absence, the audience and quality of facilitation may both wither away as well.

Off-sites. Although often misconstrued as expensive boondoggles, there are low-overhead means of conducting off-sites that won't raise eyebrows or hackles of G8 folks or congressional oversight committees. An off-site doesn't have to be sexy to be effective. Simply getting your staff away from the headquarters will provide benefit and focus enough. Perhaps it's as simple as conducting your event at the post education or warfighter center, where you can secure a classroom free of charge and meddling. Off-sites have the benefit of providing you with a captive, distraction-free audience, where phone calls, emails, and high-maintenance bosses and higher staffs can't effectively (or easily) peel away your folks. If given the latitude and proper planning, you can also enhance the overall experience by tying in a social hour, dining–

Offsite Perceptions

in, or other team-building event with it, depending on your forum and complicit bosses.

Make it a glass ball (i.e., one that can't be dropped). It is unfortunate that more quarterly or semi-annual training briefs don't enforce and include professional development tracker slides with the same vigor applied to weapons qualifications, CO_2 facilitators, or online SERE training. Take pause to think about the countless hours of generic, often mindless mandatory training you have to endure every month in an attempted flail-ex to adhere to AR 350-1 requirements. If you can shoehorn the time into your busy training schedules for such quality events, you can certainly make room for the enormously higher pay-off of sound professional development training. If push comes to shove, choose the latter over the former (almost) every time; there are few bosses out there who won't underwrite your quality efforts and accept the requisite risk.

Dual-purposed. Have no doubts that such developmental forums serve not only as great opportunities to focus on some of the keystone aspects of our profession by investing in our people, but they also pay huge dividends as team-building events, another all-too-often underrated utensil in your warfighter construction and sustainment tool kit.

Build Teams Well

WHILE SERVING as a Cavalry Squadron XO, my boss conceptualized and led a model unit team-building event several months prior to our deployment to Iraq. He chose to take advantage of our proximate Cavalry heritage at Fort Riley by employing a staff ride to the Little Big Horn. This forum would be a two-pronged approach, incorporating both professional development and team building bennies. Preparation began months in advance with the issue and mandatory reading of a book that comprehensively described the events leading up to and including the demise of Custer and his vaunted 7th Cavalry. A series of professional development sessions followed, where senior NCOs and officers were tasked to discuss different aspects of the campaign—all facets that closely aligned with their current job titles or functions.

After many weeks of preparation and context building, we finally embarked upon the staff ride. Very early one weekday, the Squadron's senior NCOs and officers piled into a tour bus for the 15-hour journey to the battlefield in South Dakota. Along with the Stetsons, spurs, and other Cav propaganda that populated the bus, three kegs of dubious contents had mysteriously been smuggled aboard, where they too would provide moral support, lubricate the occasional Irish limerick or drinking song, and help a great many of us ease the passing of those seemingly endless miles.

After a brief night's stay in a local hotel, we stumbled out upon the battlefield and followed the same path that the ill-fated 7th Cavalry had walked over 130 years earlier. Many booze-inspired headaches would accompany us along the way, as would the occasional spontaneous 'feeding of the prairie dogs,' but through the combination of sinful, indulgence-inspired agony and the lessons learned walking those hallowed grounds, relationships would flourished and teams coagulate under the hot South Dakota sun.

The staff ride was followed later that evening by a dining-in back at the hotel, complete with dress blues, grog bowl ceremonies, skits, and flip-cup tourneys that would all work towards forging the cohesive bonds that enabled our Squadron to attack its upcoming combat deployment with a surety, fervor, and esprit de corps lacking from many of our peer units.

As a line unit leader you probably devoted a significant amount of time and effort toward establishing cohesive, trusting teams imbued with a sense of identity. These team-building efforts may have included PT and athletic contests, off-duty events, old school chariot races, staff rides, hails & farewells, and professional development sessions that spanned multiple grades within your organization. So, as a staff leader, why would you ever forsake the opportunity to do the same with your ostracized and abused staff choagies? These often-perceived boondoggles are an essential element in getting to know your people, getting your people to know your people, and getting your people to know you—all in conditions that are ideally duress-free and trust provoking.

The crux... The difficulty in this effort often lies in the challenge of shutting down the staff and removing all or some of its capability from the greater unit for any length of time. Imagine the angst or frustration instigated by a sign on the front door of the Battalion HQ stating 'closed today for team building.' With the exception of the occasional lobster-eyed Soldier in those adjacent units, few people will miss the fact that Charlie Company didn't seem to be around in the motor pool today after 1300. But everyone will know and feel the impact of the absence of the entire staff for any duration.

Don't get too greedy. A full day may be a tough sell, especially for an entire Battalion or Brigade staff. A half-day event is probably the most you could hope to push for without being told that there "won't be a next time." When time is short, an extended lunch break at the bowling alley might even suffice, where you can combine a quick cheesesteak and a few words with a half-dozen gutter balls.

Protect it. To better enable such events, provide predictability for the rest of the unit by planning them well in advance, posting them on training calendars and schedules, reinforcing them at meetings, nesting them with other training events, and certainly soliciting the boss's approval and support beforehand.

Military-oriented events. Ideas for team building might include either military-oriented escapades or anything but. They can also vary drastically in overhead and preparation time. Some examples of military-friendly events (those that are certainly easier to sell to the boss) might include: Tactical Exercises without Troops (TEWTs), a quick SIMNET or CCTT battle following a hasty OPORD (fun for all branches), tactical golf cart maneuvers, a brief staff ride to a local museum or historical point of interest, an afternoon at the confidence obstacle or leader reaction course, or an arranged familiarization of another branch or service's equipment (ship tours, aircraft, howitzers, and of course, score big points by showing those dirty REMFs a tank). These are good team building opportunities with some collateral, synergistic professional development effects as well.

For more bang for the buck—or rather, the hour—use your specialty branch or service component members on staff to do the legwork and coordination, and to conduct the tour for you. It's a great way for that staff officer to plan and prepare for a training event, and to showboat their otherwise obscure branch or career field to their peers (but good luck with that Chemo!).

Non-military events. The range fans for nonmilitary-affiliated team building events are wide indeed. These might include: a simple lunch-time BBQ or potluck in the Battalion area, nine-holes at the post golf course, charter fishing trips, zip-line tours, bubble soccer, hash-runs (booze optional), or a (booze-free?) afternoon at the indoor firing range.

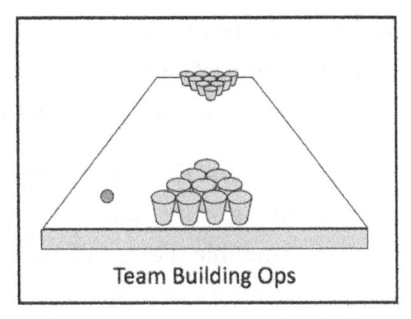

Team Building Ops

Team Think. Solving problems in groups is a simple, time-honored means of building teams. This might have happened in your younger days out on the Leadership Reaction Course, as you all collectively tried to figure out how to heave a tire over a telephone pole in some sultry training area. Perhaps this resource-intensive option is not available to you now, so let loose your imagination in tasking your staff to develop and solve problems of either a mental or physical nature. Task organize groups by breaking up those chronic cliques and creating ad hoc teams within your organization that might not have otherwise evolved without your guiding, benevolent hand. You might be surprised at the relationships that sprout as a result.

Friendly competition. Few techniques build teams better than shared physical strife, but approach this cautiously, as your efforts may backfire by inadvertently ostracizing those of limited physical prowess or athletic talent. The last thing you want to see is an over-competitive team yelling at some guy, "come on, man, hit the damn ball...you suck!" In a memorable, cohesion-inducing operation within our eclectic unit, the Commanding General lured the entire staff to the gym for several midday, murderous games of dodge ball. Teams were built that day through trial and tribulation among the red welted throngs of the uncoordinated and the pulled groins of the more seasoned among us. I saw people on staff I had rarely noticed before, and I relished occasionally hitting them upside the head with an overinflated kickball (there weren't any wrenches). Events like these are the things you will always look back upon and cherish, and the shared experiences are priceless in building unity and establishing team identity.

Timing matters. However you plan for it, I recommend conducting such events during duty hours—or partially overlapping duty hours—as much as possible. Few staff members would look forward to an after-hours Parcheesi tournament at the Leader's Club with their office-mates following a full day of nug work.

Non-military personnel considerations. If you have contractors or DA civilians working within your ranks, be cognizant of what you can legally expect to accomplish with these groups as well. Oftentimes one or both might require formal submission of leave hours to participate in such events depending on how draconian your regime is (or how loosely the contract was written). It goes without saying that this is a discouraging prospect for some. Do the math and exploit existing loopholes for the benefit of the entire team's potential. Make it interesting for these demographics and your chances of their attendance will increase exponentially.

It's a multi-faceted approach. Team building extends way beyond the occasional fun-ex during duty hours. Establishing routine hail and farewells, recognition events, town hall meetings, team huddles, sensing sessions, and a climate of mutual trust and support through a methodology of 'march to the sound of the guns' will contribute to establishing a solid identity, a sense of trust and belonging, and a reinvigorated approach to supporting one's peers and the organization writ large.

Write Evaluations, and Write Them Well

SO I KNOW THIS GUY. In one of his more recent Army assignments, his very proximate boss directed him—for the second time in a row—to draft his complete evaluation, including both rater and senior rater comments. A smart man once told him never to be humble when assessing himself in such a fashion, so he conceived a rather glowing (but accurate?) evaluation on himself, especially in the enumerated first line of both rater's comment blocks. Seventy-six days after his PCS, and only a mere two weeks before it was due to Human Resources Command (HRC), his boss wrote to tell him that his OER was ready for his signature. He logged on to review the bidding, only to find that his optimistic stratifications and cleverly selected adjectives had both been rather clumsily changed, and not necessarily in his favor. He signed the document anyway, but wrote back requesting that the OER be delayed until the 90-day mark before submission into DA, as his promotion board was convening the following week. It would undoubtedly do him no favors as it was currently written. Somehow, and perhaps a bit lost in translation, the reply came back that his senior rater was willing to consider rewriting his comment block, but insisted that the rated officer do it yet again, but this time to write it to appear 'more reasonable.' There was no mention of honoring his delay request. He wasn't sure if 'irony' was the word he was looking for, but the fact that he was being told to rewrite his own comments on himself, written on behalf of his boss and his boss's boss was, well, perplexing and disheartening to say the least. Despite the mystifying circumstances, he immediately wrote back that he would tidy up the comments and resubmit the evaluation for signatures again. Shortly thereafter, while composing the notification email to his boss that the revamped OER comments were ready for review, he received an automatic notification that his OER had been received at HRC. Apparently, and in relatively short order, patience (or give-a-crapness) had run out, and the original, hastily chopped document had been submitted, where it would go on to live in infamy much to the officer's professional distress.

Questionable leadership. As a boss, few things shout, 'I suck' more than: "write your own evaluation (or award) and send it to me for signature." If you can't set aside the time in your self-ingratiating schedule to write an accurate evaluation of your own subordinates—especially as a rater—then settle back down to the bottom of the pond, and adjust your priorities. Don't kid yourself...if that's your MO, you don't deserve to refer to yourself as a leader who "takes care of Soldiers." Those suspect leaders who find themselves in this position are typically the ones who failed to conduct initial counseling to begin with, a session that should have included a review of a subordinate's evaluation support form (job description and categorical performance objectives), among other things. Oh, but in the (hopefully) slim chance that your boss does make you write your own evaluation, don't paint yourself into a modest corner; come out swinging.

Check the block. Slightly less of a crime, but still Bush League, is to solicit your subordinates to develop and submit both 'Performance Objectives' and 'Significant Contributions' (by comprehensive—if not exhausting categories—on the latest EES-driven support form) only at the moment when their evaluation is due.

Rater considerations. As a rater, if you find yourself unable to complete your portion of a subordinate's evaluation without having to lean on the crutch of his or her support form, then you have probably failed to spend the requisite time counseling him or her over the rating period. Support forms are handy, and it does serve as a means for the rated officer/NCO to showboat their accomplishments (which hopefully match or exceed their performance goals), but if you find yourself drawing a blank without it, get out from behind your keyboard, and start talking with and counseling those deserving troopers now.

> *Proposed comments and well–crafted support forms are essential in securing the rated officer or NCO the most accurate and deserving evaluation possible.*

Senior rater considerations. As a senior rater, I would submit there's a little more room for forgiveness in this regard. It's not unusual for raters to compose 'proposed senior rater' comments for those upper–echelon leaders who don't have the privilege or opportunity of rubbing elbows with their subordinates two levels down on a daily basis. This can be especially true in circumstances where senior raters don't even reside at the same installation as the rated officer. In most cases, proposed comments and well–crafted support forms are essential in securing the rated officer or NCO the most accurate and deserving evaluation possible. If you are the rater, then caveat those submissions with a face–to–face meeting with your boss (their senior rater) to discuss your observations of subordinate performance and future potential. For those Soldiers who deserve it, you will need this time to 'sell' your boss on this potentially otherwise obscure name and face on a crew battle roster or staff manifest.

Rater Comments. When constructing officer rater comments, don't get lazy and block copy excerpts from your subordinate's digital support form into the space and declare victory. There is some debate in the community as to whether you should subtlety 'categorize' the officer with a first-line descriptive adjective/adverb ('solid performance,' 'excellent performance,' 'outstanding performance,' 'ludicrous performance,' 'hit rock bottom and kept digging,' etc). Rubbish. I recommend spending less time thinking up cryptic, flowery verbiage and more time quantifying the Soldier against his or her peers up front (e.g., 'Number 2 of 10 Captains I rate'). Less impactful and only slightly better than vague descriptive terms is the use of percentages or fractions. 'Among the top 20% of Field Grade officers' or 'top third,' etc., gives board members something to sink their teeth into, but it's a tenuous grasp at best. Expend only the smallest amount of real estate highlighting accomplishments that best impacted the organization, or by focusing on specific skills and attributes (bright, analytical, creative, etc). Be aware of what you convey to board members with a deliberate lack of enumeration. Many will only assume the worst as they try to read between the lines and piece together an image of the officer through your meandering narrative. The latest EES-driven OER format ensures the rater's focus remains on performance, while the senior rater's is directed towards potential. I think it is still important for the rater to briefly discuss when the officer should be promoted verses his peers, or sent to school (immediately, before or with peers, after further seasoning, etc.). But any mention of potential in the rater's comments will result in a quick rejection from HRC.

Senior rater comments. When constructing or drafting senior rater comments, don't feel you need to fill up the entire block. It may only serve to further frustrate a board member (although the new, automated form may help in this regard). A third to a half of the block is perfectly acceptable, especially if the writer is bold enough to rank the rated officer up front. The first line of both rater and senior rater comments is the 'beachfront property' of the write-up. Expend your words wisely here. Again, use this line to quantify the officer's potential amongst his/her peers. Dispense with the vague adjectives until later, if at all. Spend only minimal space citing specific accomplishments, and orient those on strategic or creative thinking and operational impacts if possible, as boards are really looking at this block to get a sense of the scope and potential of the officer to accomplish greater things at higher ranks, echelons, and positions of responsibility. Remember: be concise and orient on promotion, command, and schooling. Consider using those last lines to address command selection and promotion verses peers. Don't get overly dramatic or theatrical here; I would refrain from comments such as "this Lieutenant will command a Division someday!" etc.

What You Wrote:	How the Board Interpreted it:
"Number 2 of 19"	
"Top 25%"	Yawn
"Top Third"	
"Excellent Officer"	
"Solid Performer"	

Dumb it down...a little. Whether writing rater or senior rater comments, be cautious with your use of acronyms and don't assume the board members know anything about your branch or enigmatic functional area. "Dave performed admirably as the ODMFIC during our unit's recent exercise, Vindictive Llama." This sentence conveys very little to a disinterested and frankly ignorant board member who is simply trying to discern, through the white noise of your writing, where the officer stands relative to his or her peers.

NCOER comments. As there are new NCOER forms out on the market now, any of my recommendations for completing the old ones are probably rendered moot. Now that both raters and senior raters will be forced to distribute and 'block' NCOs in a fashion similar to current officer evaluations, many of the aforementioned points likewise still apply. After several years with the new forms, perhaps we can discuss some social norms in a future publication. I would, however, strongly encourage raters to utilize the support form, or perhaps the actual NCOER form itself during initial counseling, where you can both work together to determine what qualifies (quantifies?) a 'top-block' or equivalent rating for each respective area or narrative section.

Taboo comments. As convenient as it would be to blame the distribution system for the center-of-mass box check you were forced to give to one of your prized subordinates, you can't. Any mention of "I would have given this guy a top block, but my profile is maxed out" will quickly earn you a rejection notice upon submission. If you feel that strongly about your minion, let the enumeration do the talking. Board members will take note of your low population of rated officers, and won't have to read too hard between the lines in determining that top-block equivalent write-up evaluation report.

Develop a philosophy. Help yourself and your deserving subordinates early on by developing and maintaining a rating or 'blocking philosophy' that will allow you the future flexibility to appropriately recognize superior performers. This flexibility is difficult to create or maintain if you push your profile to the '49%' max authorized for above-center-mass evaluations. Abiding by a 1/3—2/3 rule or similar methodology will ensure you've created an adequate buffer to cope with the evaluations process 'fog of war.'

Develop and Manage a 'System of Systems'

MANY ARE FAMILIAR with small unit 'actions on contact' or battle drills; think through and apply a similar methodology to the development of systems to cope with information flow, administration, and critical events in your unit or on your staff. A system is defined as a combination of related parts organized into a complex whole; think 'Staff Battle Drill.' Undoubtedly, when something goes wrong, the first question a 15–6 investigating officer asks is, "what systems did your unit have in place prior to this event?"

> *"The time to fix the roof is when the sun is shining."*
>
> *–JFK*

Minimize movements to contact. Don't wait until your unit experiences trauma to figure out what to do. As JFK said, "The time to fix the roof is when the sun is shining." Think things through ahead of time, and work on unit and staff–wide 'battle drills' for both garrison and forward deployed operations to address such sobering events as suicides, ideations and attempts, off–duty accidents, training/combat–related fatalities & severe injuries, a Soldier/family member serious or violent crime, loss of sensitive items, etc. Hopefully, there is an SOP or flow-chart battle drill in place, and if so, take the time to closely scrutinize it, and ensure your staff knows exactly what to do when the time comes.

Manage routine information. Develop a system within the unit that addresses information management such as document flow to and through the command group (evaluations, chapters, awards, policy memos, ammunition requests, purchase requests, reconciliation documents, etc.). Develop and enforce an intuitive unit staffing worksheet standard for every document that hits the command group's desk; nest that standard with higher's.

Manage critical and exceptional information. Develop systems within your unit's operation centers and staff duty to manage information and knowledge efficiently and expeditiously. Treat the staff duty like a forward-deployed tactical operations center (TOC). It's certainly a more constructive use of their time than watching countless hours of cat videos on YouTube. Establish systems that manage CCIR. Ensure a staff officer has overwatch on updating their books, calling rosters, duty rosters, in/out briefs etc. This may be a responsibility discussed and shared with the CSM as well.

Wide range fans. Here are but a few additional areas that might warrant a 'systems' approach, both in and out of theater:

- Medical Evaluation Boards
- Discharge chapters
- Turning in broken computers, radios, and sensitive items
- Managing recoverables
- Unit equipment dispatching
- Services (all types)
- Ammunition forecasting
- Oil Analysis Program (AOAP) & Test, Measurement & Diagnostic Equipment (TMDE)
- 5988 processing
- Unit DTMS and training schedule approval
- Evaluation & awards processing
- Managing non-deployable Soldiers

- Communications Security (COMSEC) changeover
- Block leave procedures
- EPA spills
- FLIPL management
- Serious Incident Reports

The list is long. Pin the rose on staff officers (through initial counseling) to maintain their respective systems, and to exercise & update the SOP or policy. Ensure your systems and SOPs are nested with higher's, and use the IG's office to help validate your systems when in garrison. Solicit input from peers in adjacent units for more assistance. You should not have to reinvent the wheel; the odds are that someone smarter than you has already developed a proven system for much of what's been discussed above.

Rehearse, rehearse, rehearse. Proof your systems with notice and no-notice rehearsals.

Liaison Officers: Required but not Authorized

CHOOSE WISELY. Don't be too quick to make your token wall-hugger into the higher headquarters' liaison officer. Many units will view the inevitable, untimely, yet certain LNO taskers from higher with disdain, and rightfully so based on our meager personnel availability (and it will only get worse). This is especially true in light of the fact that the slot is pulled out of hide by committing an MTOE or TDA assigned officer or NCO to this duty from a unit that is most likely already coveting every precious head. Within these bleak constraints, commanders may utilize the LNO tasker as an opportunity to unload some 'dead weight' on the higher staff, as opposed to cutting deeper into the lean, competent flesh of the unit's talent. So, it may be your intent to stick it to The Man by providing that mouth-breather to simply answer the phone at your higher headquarters' CP. However, I would submit that investing a top-rate trooper into a liaison capacity instead, will pay off in hefty measure, and can impact your unit in ways a Soldier of equal rank or competence could rarely hope to achieve while remaining under your colors.

Combat multiplication. The right LNO can truly execute an economy of force mission that can result in precious time saved across your entire staff, by attaching key collaborative and informed planning capabilities right to the hip of your higher's staff. This will mitigate the resource-intensive effort of assuaging the ever-increasing information appetite of that higher headquarters by providing a competent Soldier that can, on behalf of your unit, provide current data, SITREPS, knowledge, unit capabilities and expertise to that headquarters Tactical Operations Center (TOC) and planning table.

> *The right LNO can truly execute an economy of force mission.*

So, the next time higher's plan calls for your unit to execute the maneuver equivalent of a Triple Lindy, you can only blame yourself and that fish-eyed LNO, who sat in the corner of the Brigade plans Drash, dozing and bobbing like a drinking bird throughout the MDMP. Be cautious, however, if that headquarters becomes too enamored with your LNO, you might not get him or her back when the show's over (true story).

There can be only one. We could only hope. Unfortunately the truth is far from it. An LNO tasker from higher probably implies that it's a 24/7 requirement. As a result, it's never just one LNO. Every position during sustained operations requires at least two shifts, thus two competent officers or NCOs must be used to feed higher's personnel meat grinder.

All this, and right on top of that recent and unjust harvest, where they shanghaied your best Soldiers to serve on the Brigade Commander's Personal Security Detachment. You bastards.

Make Time to Train the Staff on the Military Decision Making Process (MDMP)

THINGS HAVE CHANGED. You will find that many Career Course graduates and junior officers will arrive at your staff with little to no knowledge or experience in MDMP, especially at the Battalion level. In addition, our prolonged experience in counterinsurgencies sometimes allows our MDMP and synchronizing skills to atrophy, especially in relation to major combat operations. Historically, I have found that only about 30–40% of Majors in staff groups at ILE had formal experience participating in the operations process or MDMP, though by no fault of their own. Those warfighters have been kicking in doors, clearing routes and hilltops, and training foreign security forces for well over a decade and a half now. Be cognizant of the trade-offs. The time to train your staff is not when MCTP (formerly BCTP) shows up on your doorstep. Make and take the time to train the staff on MDMP out of contact and well before centralized Command Post Exercises, CTC rotations, and deployments.

Find solitude. I recommend utilizing the installation Simulations or Warfighter centers. There you can fall in on existing TOCs, CCTTs, hard/software systems, expertise, and classrooms—all away from the distractions of the Battalion HQ and the gravitational pull of email. Spend low-stress time here reviewing, training, and experiencing MDMP and developing/refining planning SOPs and staff estimates. When you feel more comfortable with the staff, nest the training with Command Post setups in the HQ backyard, a great place to evaluate setup, floor plans, jump, and tear down, refine both digital and analog systems, and rehearse outside of the scrutinizing gaze of Higher. When you feel really comfortable, invite the commander to participate. When he/she feels comfortable, then strike out for 'design integration'!

Make it happen. This sounds good on paper, but the dizzying array of colors, shapes, and micro-fonts that will be your training calendar may discourage you from attempting this. Don't let it. Bring it to life by working it in to your Quarterly Training Guidance and/or nest it with other unit training events like gunneries or Platoon Situational Training Exercises.

Those Bastards Down at Company

YOU MIGHT HAVE HEARD that 'bastards up at Platoon' joke, but don't be surprised when you and your staff start thinking along the opposite lines, 'those bastards down at Company.' You will be astonished at how the space between the Battalion HQ and the Company CPs starts to look more and more like a DMZ over time. Tension between the staff and companies is common, but if not checked, it can be debilitating to the unit. Here are some common contributors:

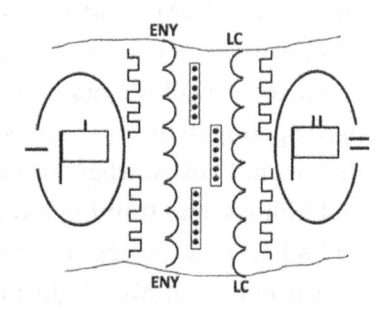

Sticks-in-the-eye. IPRs and meetings devolve into dime-dropping sessions as staff officers try to leverage companies into compliance with due-outs or FRAGOs. Nothing energizes Company Commanders like bold, red-filled bubbles or 'delinquent' next to their unit call sign on a Command and Staff slide. This technique has its place, but never use it unannounced. Ensure the dirty laundry has been aired out ahead of time, and subordinate unit attendees know what flaming turds may be floating their way when they step foot in the conference room. That way they can come prepared to 'alibi' or discuss a 'way ahead' without being put on the spot in front of their boss.

Battalion Deployment Prep Tracker			
Unit	Milestone A	Milestone B	Milestone C
A Co	✓	✓	✓
B Co	✓	✓	✓
C Co	🗑	🪳	💩

Double-tapping Companies (dipping multiple staff soda-straws into companies). The following scenario is not uncommon: the Chemo walks over to the companies and asks the First Sergeants for this month's USR data by COB Friday; the S1 subsequently calls the Company training clerk and wants the requirements for personnel shortages submitted by COB Friday; the S4 calls the Company XOs and demands the top ten equipment shortage list by COB Friday; the S3 calls commanders for an updated crew roster by COB Friday; the BN XO calls commanders and personally reiterates the importance of submitting personnel and equipment data by COB Friday. You can see how this might quickly frustrate a Company Commander and 1SG. If the staff requires data or products from subordinate units, ensure it gets consolidated, proofed, and issued out in the Daily FRAGO. No other medium should be tolerated. That way the S3 shop can maintain a highly visible, on-line/SharePoint, and updated 'FRAGO compliance—suspense tracker' spreadsheet so that there are no surprises for anyone.

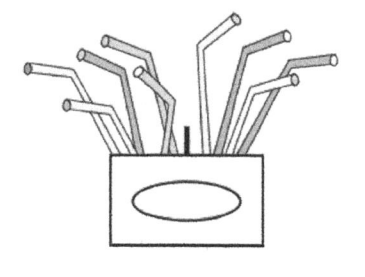

'Take a number, First Sergeant.' Staff sections don't maintain a 'customer service' oriented attitude when dealing with companies. Sometimes staff officers start to think that their affiliation and proximity with the Battalion headquarters means that companies answer to them and not the other way around. The Battalion Commander will have a big say in the role of the staff in relation to subordinates, and it will fall on you to enforce his/her intent. Many prefer that staffs support both the Battalion Commander and Company Commanders (in that order). Staff guys are busy, and it will be a challenge for you to instill a 'drop everything' mentality on a staff section when a 1SG or Company Commander needs a hand. If it's not in conflict with something the Battalion Commander wants, then they may need to. If you can breed that kind of service-oriented philosophy, the sense of unity of effort and mutual support between staff and companies will reap huge benefits in more important ways.

A sensation of hostilities. When units continually fail to meet suspenses and deadlines, utilize Escalation of Force (EOF) procedures. Don't run into the Battalion Commander's office complaining the moment A Troop fails to turn in their training schedules on time. Start with your current operations/battle Captain making the reminder call. Still no response? Maybe it's time for the operations Sergeant Major to work over a First Sergeant. Nothing yet? Now it might take a Major to get on the phone with a Company Commander. Up the ante once more by paying an office visit. Be cautious though; when that happens too often, Company Commanders may become desensitized to your nagging, so pick and choose carefully when you need to weigh in. When it becomes glaringly obvious that a commander is just not pulling his/her weight or deliberately blowing you off, then ask your commander for help. Come armed with facts and trend analysis, and maybe even a draft email. You or the Operations Sergeant Major may also choose to approach the Command Sergeant Major with the issue, and let him/her work it through the First Sergeants in a less visible yet highly effective fashion.

Those Bastards Up at Brigade

IT REALLY IS A GRUDGE-FREE ZONE. Your higher headquarters will undoubtedly serve as one of the greatest sources of your daily frustrations. The FRAGOs, taskings, GFIs, taskings, no-notice meetings, taskings, immediate RFIs, etc., may get you thinking after a while that Brigade has it out for us: "they just don't like the Cav"; "they're biased against sustainers"; "don't trust us FA guys"; "if you're not an Infantry Battalion, they don't care about you..." Pick your complex. Just remember, those folks that work in the various sweatshops up at Brigade are most likely your peers and former classmates. They too are good Americans, and don't wake up in the morning asking themselves, "How can I stick it to the support Battalion today?" They are fallible, toiling in the salt mines, trying to manage resources and solve problems. They don't have time to develop and nurture a grudge against your unit, so don't take it personally, and don't pole vault over mouse turds.

Exploit the 'Major Mafia.' Work hard to build trust with Brigade or your higher HQ staff. Don't dime them out at meetings without prior warning. Most issues can be resolved outside of public forums. Don't be the guy who sharpshoots

Brigade's Unbiased Tasker Wheel

Brigade staff officers and their products in meetings in front of their bosses. When you need that guy or girl to do something for you in a pinch (like update a bad stat on a command & staff slide minutes before the Brigade meeting), they may not be there for you. Ultimately, your unit suffers. Remember, these guys are your peers or work for your peers. If you earn a reputation as a sharpshooter or backstabber with your peers on that higher staff, you will hurt yourself and by extension hurt your unit. The Brigade staff has the ear of the BCT Commander... he or she will learn of your reputation from his or her staff. Keep them on your team for the good of all.

> *If you earn a reputation as a sharpshooter or backstabber with your peers on that higher staff, you will hurt yourself and by extension hurt your unit.*

Chapter IV

The Calculus of Field Gradeship

WHY CALCULUS YOU ASK? It was always such an intimidating subject in both high school and college; the ultimate and mysterious white citadel of mathematical accomplishment for the numbers–challenged, nervously flatulating, liberal arts majors among us. What better scary-sounding, daunting term to preface a chapter that discusses the essential facets of the Field Grade crucible? Science assumes a peripheral role at best in this chapter, as we will dig deeply into the art–heavy tasks of relationships, strategic thinking, understanding the 'operational environment' and shaping it for the betterment of our units, Soldiers, and staffs.

Iron Relationships or Bust

THEY MATTER...A LOT. Don't think that by sheer rank and intimidation that you will be able to bull your way through the 'Iron Jobs' to success. You need to solicit buy-in, loyalty, and trust from up, down, left & right, and beyond. Your influence in and outside of your unit will have a direct correlation to your success as an S3 or XO, and ultimately, to the unit's.

> *You need to solicit buy–in, loyalty, and trust from up, down, left & right, and beyond.*

Peers. One of the most challenging relationships is often with your fellow Battalion Field Grade(s).

Working within the commander's guidance, sit down early with your peer and iron out potential friction points. Establish your respective boundaries and domains, and don't assume that doctrine clearly dictates what the two of you should do. The following are some areas that might need further clarification or discussion:

- Battle rhythm management
- Tactical operations center (TOC) operations
- MDMP staff and timeline management
- Operational mission command
- Operations Sergeant Major management
- USR & Chemo management

- Taskings for investigations
- S2 command & control (in and out of the field/theater)
- ISR planning & synchronizing
- Fusion cell operations
- Staff huddle times and attendees
- Command maintenance attendees
- Battle/Commander's Update & Assessment (BUA/CUA) prep & conduct
- IPR chairs
- Staff PT
- Event scheduling (services, inventories, etc.)

The two of you may not see eye-to-eye on every issue and that's fine. Don't feel that you both need to speak as one to the commander. This limits options for the boss and creates groupthink. If you are going to recommend a different COA than your counterpart, however, give the other the courtesy that you are going to do just that, so your peer does not think you are trying to backdoor him or her. Maintaining trust in the competitive world of Iron Majors in key and developmental (KD) jobs is essential to providing your organization with professional, drama-free senior leadership.

Although it helps for the two of you to be friends, it's not necessary. You do, however, need to respect and support each other. If the staff or others in the unit determine that there is a rift between the two of you, expect them to exploit the seam with 'play mommy against daddy' games.

Don't make the boss be the referee or marriage counselor between the two Field Grades. He or she has enough personnel friction to deal with, so don't waste the commander's time with Field Grade relationship issues.

Don't forget to cross-train roles, as you will most likely have to assume the responsibilities of the other for undetermined periods of time (leave, EML, PCS, transitions, etc.).

Senior NCOs. Within the organization, build trusting relationships with the Operations Sergeant Major and First Sergeants. They hold the informational reins of the unit. They, more than any others, understand the impact of your planning & GFIs. Their feedback and assistance can be invaluable. Try to throw your bulk around these folks, and you'll find yourself out in the cold, isolated and ineffective. The operations Sergeant Major may get handed from Major to Major when the unit is deployed, in the field, or back in garrison. Between you and your peer, figure out how to minimize the impact of the Sergeant Major shuffle, so you both can maintain open lines of communication without overburdening him or her.

The Command Sergeant Major. Get your boss's intent on your relationship with the Command Sergeant Major. Divvy up responsibilities early to prevent clashes or underlaps. Subjects worthy of discussion might include:

- Non-deployable rosters, reporting, and management
- Chapter management
- MEB/MMRB tracking
- MEDPROS issues/tracking
- Staff NCO moves
- Staff duty issues and procedures
- Ceremony planning & resourcing
- Calendar impact events (NCODP, Soldier of the Month/promotion board, EIB, EFMB, etc.)

Then, meet with the command Sergeant Major at his or her discretion and work out the details.

Be there for Company Commanders. You may find that, consciously or subconsciously, your actions and behaviors will serve as one of the biggest influencers for those senior Captains, as they start to look past Company command in determining their future destiny in or out of the service. Be approachable; coach, teach and informally mentor. Provide top cover and set yourself up to be a 'GFI sounding board' for Company Commanders. Protect them from themselves. You may have to be the 'bad cop' with them at times, but make sure they always feel they can still come to you for advice and guidance. Your measure of performance in this regard can easily be quantified by the frequency in which Company Commanders visit your office.

Reinforce your flanks. Outside of the organization get to know and share information and ideas with Field Grades in adjacent units, the support operations shop, and on Brigade staff. Don't hoard ideas. Foster a constructive, sharing network with your peers, and it will reap huge benefits and save time. Share and ask for things like SOPs, meeting formats, policy memos, OPORD formats, cool kit/part national stock numbers (NSNs), TTPs, example investigations, etc. Typically the Brigade Commander places as much weight on your ability to work with others as he or she does on how strong you are as an individual performer. So for the good of all, embrace the broader team concept (more on this later).

Make a friend at JAG/Legal. He or she is in a prime position to assist your unit with chapters, FLIPLs, 15–6s, and other investigations. Borrow and include your legal clerk in Battalion command & staff meetings to brief chapter statuses and UCMJ.

Outside organizations. Get to know the Department of the Army civilians and others that work at or run agencies on post. Consider the following key facilities and functions:

- Directorate of Logistics (DOL)
- Department of Information Management (DOIM)
- Installation Travel Office (ITO)
- Training Aides and Support Center (TASC)
- Ammunition Supply Point (ASP)
- Simulation (SIM) Center
- Army Materiel Command (AMC)
- Range Control/Services
- Morale, Welfare, and Recreation (MWR)
- Soldier Readiness Processing (SRP) site
- Central Issue Facility (CIF)
- Safety Office
- Environmental Protection Agency (EPA) rep
- Hospital Medical Evaluation Board manager
- Wash rack
- Property Book Office (PBO)

Many of these folks can be great allies and help you in a pinch when the fog of war/training kicks in. Treat them with respect, award them coins and certificates of appreciation on occasion, and recognize their efforts to their bosses, and you may have earned your unit a friend and ally for life.

The intimidator. Don't be afraid of the Inspectors General (IGs) office. It may sound trite, but they really are "here to help." With your commander's knowledge and approval, take advantage of their Remission or Staff

Assistance Visits (RAV/SAVs) and other courtesy inspections to help you stand-up or correct unit systems, especially during Reset. Many IG offices will have a huge database of SOPs for just about any system you can think of, from maintenance SOPs to ammunition management. Don't wait until the Division-mandated Command Inspection Program (CIP) to figure out (along with your Brigade and Division Commanders) that your systems are defunct or non-existent.

Take Care of the Boss

Enable the visionary. When you take care of the boss, you take care of the unit. Fight the close fight and run the daily operations of the Battalion so your boss can look at the big picture, fight the deep fight, and make out with the good idea fairy. Remember, however, that the close fight is not the knife fight. Let your staff do that so you can spend unobstructed time thinking about, developing, and refining your systems.

Secure the flanks. Ensure your boss has the most accurate and up-to-date information before he/she steps into his/her next meeting at Brigade (command and staff, USR, CUAs & BUAs, OPORD backbrief/confirmation briefs, etc.). When your boss gets chewed on, it's probably due to something you failed to do to set him/her up for success. You should take it personally and adjust accordingly.

Establish a screen line. Company Commanders get to ride the Green Tab Express to the Battalion Commander's office, but few others should. Don't let your staff officers or other random choagies approach the commander without having seen you or the adjutant first (preferably in reverse order). That is, of course, unless they are exercising his open door policy to complain about you.

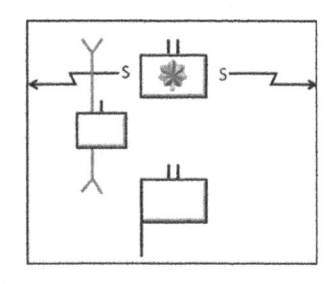

Horrible bosses. So what happens in the rare event that you find yourself working for some tyrannical, selfish boss that, despite your best efforts, you just can't seem to figure out how to like or respect beyond the silver oak leaf or eagle? If that unlucky guy is you, then you'll just have to get over it for the benefit of the unit. You may find yourself having to run more interference for subordinate units and staff sections as a result, but making your boss look good will have a positive impact on your unit and its Soldiers. Regardless of what you perceive the commander's motives to be, dubious or otherwise, or whether you agree with them, your support for the commander will ultimately translate to better outcomes for the unit.

Exit strategery. Influence your boss to go home by 1800, for his/her own sanity...and yours.

Working with the Boss

BOSS IPB. Try to understand how the boss thinks and interprets data. Determine how the boss processes information; is he/she a visual, verbal, or text learner? Is he/she at optimal efficiency in the morning, afternoon, or evening? Does the boss understand the Operational Environment through meetings and dialogue, or through read-ahead packets, decision memos, and post-meeting reflections? Comprehending how the boss thinks and when he or she thinks best will better enable you to

> *Comprehending how the boss thinks and when he or she thinks best will better enable you to package and present information that will result in clearer commander's guidance and intent.*

package and present information that will result in clearer commander's guidance and intent.

"What's your sign?" Don't be afraid to tell the boss how you think, learn, and process as well. This may enable him/her to issue more understandable guidance to you. You may be at opposite corners of the Myers-Briggs Type Indicator matrix, but it doesn't need to inhibit shared understanding between the two of you.

Maintain the Decision Support Matrix (DSM). Track decisions facing the commander, both internal to the organization and beyond. Recall that one of the fundamental purposes of the staff is to manage and analyze the bushels of information that enter the unit from all directions on a daily basis, and reduce it down to digestible chunks that will enable the commander first to understand, then to make informed decisions. Comprehend and define the milestones and timelines that support his or her decisions. You may want to consider a 'lines of effort' or DSM approach, which visualizes in time and space the potential decision points the commander needs to make by critical event, what information is required to make the decision, and when they need to be made. Also understand your higher headquarters decisions and their potential implications on your boss and unit as well.

Support by fire. Back the boss's decisions, for better or for worse (within moral and legal range fans, of course). Beware of the ego-facilitated pitfall of doing the contrary, whether overtly or subconsciously. The boss may prompt you for your opinion as to which of two courses of action you think is best. Despite your vehement loathing for the other COA, and the logical and eloquent case you may have presented in favor of yours,

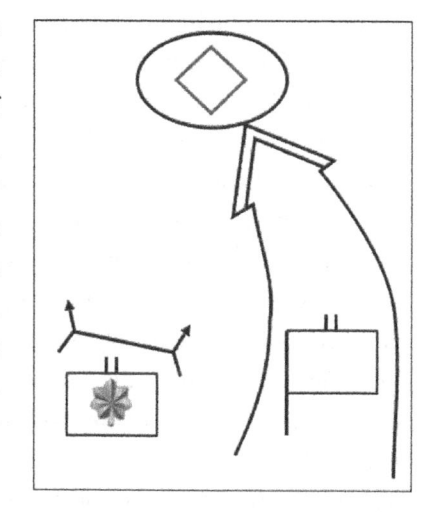

CHAPTER IV: THE CALCULUS OF FIELD GRADESHIP 121

don't be personally offended when your recommendation or COA is rejected in favor of another. When the boss finally decides to pursue a COA, whether it's the one you advised or not, you now have the often difficult obligation of owning it yourself, lock, stock and barrel. You must dedicate yourself and the staff completely to its execution. In the dark recesses of your brain where your bruised ego goes to lick its wounds, you may secretly want the boss's COA to fail just to prove that you were right to begin with. This malicious desire may induce skew on your ability or that of the staff's to accomplish the mission according to the commander's intent. The staff can't be allowed to detect an ounce of doubt or cynicism on your part, as they will be that much more disinclined to support it themselves. This can be a real challenge for some; don't let it be for you.

The Inner Circle of Trust. As a Battalion Field Grade, you gain automatic entry into the "Enclave of the Big Five" (CDR, CSM, SGM, XO, and S3). It does not, however, grant you membership and immediate access into the more prestigious and coveted "Inner Circle of Trust." This may take some time, as trust requires and often depends on many variables, mostly centered on your boss's personal preferences and your ability to listen and provide sound and candid advice when called upon. Many bosses will look to someone like the XO or S3 to vent, rant, and rave about higher, peer units, subordinate leaders, silly policies, or even personal problems. They may look to someone to bounce ideas off of (some of dubious quality no doubt). Those select few sounding boards constitute the members of the "Inner Circle of Trust." Once mutual trust is established between you and the boss, you need to respect the Vegas ROE and the "Cardinal Rule of Privileged Information," also known as "what's said in the boss's office stays in the boss's office, unless specifically told to do otherwise (the same goes for email)." If you have any doubts as to the public release of information discussed between you and the commander, seek immediate clarification. When you violate this rule, you may find yourself, in the words of COL Mark McKnight, "voted off the island." Welcome to the purgatory of irrelevance.

Pick Your Battles Wisely

BUILD CAPITAL. When the third tasking of the day from the Brigade S3 hits your inbox, you might feel the urge to grab the phone and give him/her an earful. When your companies get split down the middle on multiple deployment flights while your sister units deploy intact, you might feel inclined to call the Brigade XO and ask "WTF?" As an Iron Major, you will probably find yourself frustrated or angered by higher on a daily, if not hourly basis. And with each additional frustration, the urge to gripe becomes almost irresistible. But before you fly off the handle in an attempt to protect your Soldiers and unit from perceived

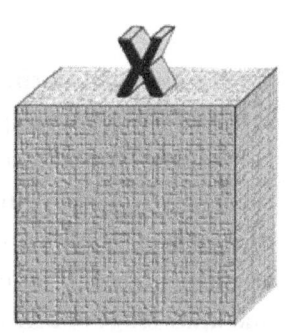

Resistance is futile?

injustice, apply some careful scrutiny. Do the fact checking and trend analysis and figure out if your argument is worth the 'currency' you may have to spend on it to get your way. You don't want to gain the reputation in the Brigade as the constant complainer or whiner. Ultimately, higher headquarters will become desensitized to your appeals, and when it really counts, your argument will fall flat. Pick and choose carefully the fights you want to fight. Saying 'yes' 9 of 10 times grants you the currency, and probably the desired outcome, when you finally do need to say 'no.'

Expend ammo wisely. To get your way with higher, you may choose to fire a Silver Bullet in the form of getting your boss involved. Ensure you do your homework first, and have all the facts before your commander gets on the phone with his boss and tries to lay out a half–baked or unjustified case. You will both loose Wasta if such an event unfolds and your boss will be less inclined to get involved on your behalf again.

> *Saying 'yes' 9 of 10 times grants you the currency, and probably the desired outcome, when you finally do need to say 'no.'*

Curb unwanted enthusiasm. Sometimes your boss may choose to get involved without your prompting. "I'm a Battalion Commander, by God. No Brigade staff officer is going to tell me what I can't do! Hand me the phone, XO!" Assist your boss in this endeavor. You may have to approach him without challenging his authority in public, and do some joint risk assessment on the 'fight' you are about to embark on. Use your expertise and knowledge of Brigade staff to convince him of a softer approach and to allow you to skin the cat a different way as long as his intent is still met.

Learn How to Constructively Advise

LIKE THE COMMAND SERGEANT MAJOR, the Battalion Field Grade can serve as an advisor to the commander. Good advising is somewhat of an art. It may be a challenge at times to convey your point to the commander without putting him/her on the defensive. This requires careful 'METT-TC' analysis to determine how and when your boss wants or even needs advice.

> *You all owe it to the boss to think through problems from multiple perspectives.*

Be (mostly) subtle. Don't be in a rush to thrust your opinion upon the boss, but when asked, be candid and tactful—a carefully balanced combination. If you feel that you are obligated to share your unsolicited thoughts with the boss for the betterment of the unit, seek counsel with your peer and the Command Sergeant Major first to get their thoughts. If they have similar opinions, then you may be on to something. Sometimes, getting the boss to ask you what you think about an issue may require you to first ask the boss the question, then wait for the counter-question.

Don't fall victim to groupthink. When asked, tell the boss what you really think whether it's in line with the others or not (tread lightly though). You all owe it to the boss to think through problems from multiple perspectives.

CAV, Baby!

Or 'Coordinate, Anticipate, and Verify' for short. This was one of my boss's favorite acronyms, and one that I failed to recall often enough. These terms exist in many facets of formal problem solving methods, and this quick rule of thumb can be applied to just about everything from planning and executing a Battalion ball to preparing for a deliberate attack.

> *What separates great Field Grades from good Field Grades is their ability to master these anticipatory skills.*

Most are familiar with, and confident in, their ability to coordinate for an operation or event, but the art of 'Anticipate' and the science of 'Verify' are sometimes overlooked. 'Anticipate' requires some careful analysis and brainstorming and may result in additional branch plans. They are, in essence, the 'what if's' of an operation. Some examples:

- "What happens if the wireless mic goes out during the guest speaker's pitch?"

- "If the busses don't show for range transportation, what's the back-up?"

- "What's the plan for fueling the TOC generators if the LOGPAC gets interdicted on the way to the JSS?"

Yes, it can be a slippery slope, and only your experience can guide you as to whether or not you have anticipated enough realistic 'frictions of war' for a particular event. As we all know, no plan survives first contact with the enemy, both figuratively and literally, so do the legwork ahead of time to cope with the inevitability of change.

The discriminator. What separates great Field Grades from good Field Grades is their ability to master these anticipatory skills. The ability to anticipate and fix problems before they happen is why Field Grade officers are paid the big bucks. The key to this is time to think. Get yourself out of the knife fight early and often. Delegate and hold your staff to high standards so you can build a level of trust and confidence in them that allows you to decentralize taskings, and grants you the space and time to ask the "what if" Spend your time anticipating what could go wrong, and then take steps to avoid failure.

Verify. The last step—verify—is perhaps the simplest but most neglected. Verify doesn't mean micromanage. It may mean simply that you have to verify that subordinate staff members have verified. Examples: verify through a General's Aide that they will or will not be attending your Battalion ceremony. Verify that the contractor will, in fact, provide the crane (and backup?) for tonight's T-wall emplacement mission. Verify that the medics will be at the range on time with the proper evacuation vehicle. Verify that your staff got the word about bringing in their pro-masks for inventory on Monday, etc. There is a lot of overlap with 'verify' and the management process of 'follow-up.' Master both and you will effectively put Murphy back in his cage.

An action passed is not an action complete. The days of saying "I told so-and-so to do it" or "Sir, we coordinated with TMP, and they said they never saw the request" are over. As Field Grades, the only thing that matters is results. The buck stops with you. If a ceremony flops, a training event is cancelled, or a suspense is missed, there is no one to blame but you. If you took the CAV concept to the fullest extent and Murphy still shows up, your boss may understand, but short of that, there is no excuse.

AAR Everything

IT WAS DAY TWO IN THE MANEUVER BOX at the National Training Center when our Brigade hosted its first Battle Update & Assessment (BUA) over the Command Post of the Future (CPOF; yep, still sounds silly). It had been a tough couple of days for my Squadron thus far, and our Cavalry Troops had encountered plenty of frustration and resistance in their simulated areas of operations (AOs), all in preparation for our upcoming deployment to Iraq.

After the Brigade intelligence officer finished his portion of the brief, the Battalion Commanders began their assessments. First in the queue was one of two Combined Arms Battalions (CABs). In a rather résumé-building fashion, the commander proceeded to review the significant actions of the day, the enemy body count, and other various triumphs his unit was able to achieve over the past 48 hours. The other CAB Commander came next, and, in true competitive if not dramatic form, reviewed all of his milestones accomplished, further damage done to the enemy, and touted other various achievements his expertly-led Battalion was able to realize amongst the player populations and upon the OPFOR. My Cavalry Squadron followed, and in proper sequence, my boss began by reviewing the SIGACTs in our AO. But then, contrary to the boasting and one-upmanship of the previous briefers, he proceeded to explain the hard lessons learned through several of the tough actions that had occurred since the 'fight' had begun, most of which had been fought to our detriment. There were no OPFOR body counts, no lists of glorious objectives seized, and no pontifications about the superb performance of the Squadron. To the contrary, he used his briefing time as an opportunity to explore what he or we should have done differently to better protect our force and achieve improved effects across the battlefield. I looked at him nervously, waiting for him to snap out of it and jump back on the celebrated body-count bandwagon, but he never did. After his brief, the silence was palpable, almost unnerving. I half-suspected that we had dropped off the net when the Brigade Commander finally came back on, speaking deliberately and emphatically: "now that is what this update should be all about!"

Through his own humility, my Squadron Commander was able to affect a paradigm shift on that day that made the Mission Readiness Exercise less about what he or the unit was able to achieve, and more about what they were unable to achieve, and why. And, at a training event such as this, would you really want to have it any other way?

That day's brief set a precedent that would span the rest of the rotation. But of far greater importance, it's essence would persist through our combat deployment several months later, establishing those Battle Updates, assessments, and other such meetings as critical opportunities for cross-talk, shared lessons learned, and further introspection and growth, that would enable this fighting BCT to adapt more rapidly to a shrewd enemy operating within an extremely complex AOR.

One of the Army's greatest cultural norms is our ability to frankly assess our performance as individuals and teams. Complacency often comes with our inability to look objectively at our own systems, SOPs, and tactics, techniques, and procedures (TTPs). Applying a no-holds barred After Action Review philosophy to everything you do will enable you to maintain current and effective systems across all facets of the organization.

AARs: the basics. Don't pet yourselves too long on what went well. Note it and move on to the dirty stuff. Spend a minimum of 75% of AAR time analyzing what went wrong and how best we can fix it for next time. Like an AA meeting, the healing can't begin until we admit our problem. This requires humility on everyone's part, especially leaders. Leaders who aren't afraid to stand up and say, "my unit was destroyed in the assembly area today because I failed to do ____" inspire subordinate leaders and Soldiers at every echelon to conduct constructive introspection as well. This creates a synergistic, multi-echeloned effort to identify and fix mistakes across the depth and breadth of an organization. I'm convinced that sometimes our biggest obstacle to learning is our own ego. Swallow your pride for the greater benefit, growth, and—potentially—the survival of the unit.

> *"It is not the strongest of the species that survives, nor the most intelligent. It is the one that is the most adaptable to change."*
>
> *–Charles Darwin*

None of us is as smart as all of us (mostly). Good leaders and leaders of AARs do less dictating and more soliciting, "how do we fix this problem, staff?" Soliciting ideas from subordinates engenders buy-in and dedication that may surpass your ability to inspire them with your own perceived brilliance. You probably already know the answer to the question, but clamp your soup-coolers and lead your organization on a voyage of self-discovery. You will be pleasantly surprised with the imagination and ingenuity of your subordinate's unconstrained critical thinking.

Spend time AARing your systems. Relook your battle rhythm and weekly meetings. Spend ten minutes after a Command & Staff and ask "how can we make this meeting better or more effective? What slides need to stay or go? How can I or this staff better set you Company Commanders up for success?" When your unit has to contend with a critical event, pull in the staff and subordinate commanders and AAR the information flow, how to avoid similar problems in the future, what SOPs and policies need to be implemented or changed to cope with it, etc. Consider the following important areas to AAR:

- Suicide attempts or ideations
- POV accidents
- Training or motor pool accidents
- Negligent discharges (NDs)
- Escalation of force (EoF) incidents
- Losses of sensitive items
- DUIs
- Services & maintenance parts flow
- Ceremonies and social events, etc.

Don't underestimate the power of an AAR in combat. I would submit that a unit that aggressively incorporates AARs in combat is, in essence, a learning organization, and one that will quickly adapt to stay well forward of an enemy's decision cycle. An AAR goes far beyond a storyboard, commander's inquiry, SIR, or 15-6 investigation. Looking objectively at how an enemy attacked us or vice versa—successfully or otherwise—and how we/they reacted, is an inherent step of IPB, and will allow us to create or modify our COAs and TTPs and/or equipment for the betterment of the unit.

Investigations as AARs. Like most, you may balk at the premise that investigations can be of enormous benefit to units and Soldiers. You probably think of them as added burdens and unnecessary work that detracts from the current mission, etc. You might even be inclined to nominate underachiever investigators as an 'economy of force' mission of sorts to preserve combat power. If this be you, then I would submit that you are passing up on a potent opportunity for learning and growth. If done correctly, an investigation can serve as a powerful After Action Review. It should uncover the fundamental root causes of our mistakes—not to merely assign blame or name a scapegoat—but rather to provide the unit with recommendations as to how we can prevent from making those same mistakes again. It can be a frank, humbling, and sometimes invasive look into our systems or lack thereof. Don't be afraid of investigations, and embrace and resource them as tools to make your units better.

Propagate good ideas. Don't forget to share your combat lessons learned with others outside of the organization. It may save Soldiers' lives. If your unit develops sound TTPs or (authorized?) equipment modifications that exploit enemy weaknesses, ensure the word gets out to the rest of the Brigade, and perhaps, if warranted, to the rest of the theater/Army. Asymmetric Warfare Group (AWG), the Center for Army Lessons Learned (CALL), MilSuite, branch or other professional journal publications, and even good storyboards can be great forums for getting the word out quickly. As always, manage sensitive or classified information appropriately in this endeavor.

Echo in eternity. At the end of the day, AAR observations in any environment need to be codified for future generations through the use of a formal system like a lessons learned program. This legacy database of things that went well and not so well should be the starting point for every subsequent iteration that you and your unit embark upon. Make it a habit of prefacing an exercise IPR or NTC rotation planning session with a 'lessons learned last time' slide. This invaluable collection of past generational pitfalls should become a standard launching point for your unit's future problem solving or project management endeavors.

On Occasion, Take a Knee... All of YOU

WHEN THE TRAINING SCHEDULE allows it, take leave/pass and get away from work. Take advantage of block or 'max leave' periods that appear on your training calendar as well. If you think 60 days of saved up leave is a badge of honor, then you need to readjust your priorities. You, your family, and possibly your subordinates may be paying the price in more subtle ways for your lack of vacation time. Remember that your subordinates—the Lieutenants and Captains—are making life and career decisions daily, and are looking at you and asking themselves 'is that what I want to be in five to ten years?' Set a good example, and go to the beach once in a while.

Don't just burn it. "You lost 10 days of leave this fiscal year? That sucks. Well, get cracking on those slides, slacker." You should work damn hard to avoid ever saying this to one of your troopers. It is a tragic sin to lose leave through a failed 'use or lose' effort, even as an Iron Major. But it is tremendously more so if you allow your subordinates to do the same. Manage their days carefully to ensure they don't arbitrarily sacrifice well-earned and well-deserved leave, especially in a perceived noble but misguided effort to show their commitment and dedication to the unit.

PCS it. Sometimes it can be frustrating to have a new officer or NCO arrive to take over a key billet, get their feet up under them, then have to take a month off due to use or lose. In order to prevent this kind of extended absence, ask that Soldier how much leave they are in jeopardy of losing when they first reach out to you through that introductory letter or phone call before actually signing in. You are probably chomping at the bit to get that slot filled quickly, but have that Soldier maximize leave prior to arrival in order to mitigate an extended absence once or after they arrive, and well before you start to rely heavily upon them.

Pre-prepare. Before embarking upon that cross country family road trip to visit the second largest ball of twine on the face of the earth, consider these simple tasks to ensure things don't pull apart at the seams in your absence:

- Plan your escape carefully. If your fellow Field Grade is going to assume the reigns in your absence, then good on him. This can be a stout burden to lay upon the shoulders of one who is already bearing a large load however, so consider nominating another staff officer to step-up instead. Bring your designated and trusted 2IC (second-in-charge) to the prior week's key events and meetings so he or she has some context before they assume your chair and responsibilities at the next iteration (Command and Staff, Training meetings, staff huddles, IPRs, etc.). In order to avoid the 'substitute teacher effect' in your absence, determine what monthly or quarterly events that your stuckie will have to attend (USR, SATBs, etc.), and establish talking points and even draft slides for him or her to use. In addition, it is probably worthwhile to lay out what is or is not within his or her span of control to make decisions upon, or what must be pushed to you, your peer, or the boss.

- Reach back...but sparingly. When you get tired of the stress-free endeavors of managing three kids and a spouse crammed inside your family cruiser as you tour a countryside filled with giant, plastic road-side dinosaurs and wax museums, take a short halt and give your 2IC a call to see how things are going. If you were unlucky enough to qualify for a Blackberry, then ping your minion with an occasional email requesting a status update.

When you're on leave, stay out of the office as much as you can, virtual or otherwise.

While on leave, however, limit your interactions with that devilish contraption as much as possible. It's too damn easy to get sucked back into the cybernetic workplace, where you may feel inclined to pull over at the next rest stop and reformat that abysmal Semi-annual Training Brief slide, or respond to imbecilic inquiries from folks that your 2IC can handle with equal or better grace. When you're on leave, stay out of the office as much as you can, virtual or otherwise. Those around you will appreciate it.

- Out-of-office replies. Absolutely essential to a blissful escape, the simple task of setting up your Outlook out-of-office reply will ensure there is someone available to literally answer the mail and tamp out brush fires. Clearly state the dates you will be out of the office and provide the phone number and email for your back up. For privacy's sake, I would refrain from stating where you'll be; a simple "I will be out of the office from __ to __" should suffice just fine. Select the correct option to ensure it only gets sent out to each solicitor once, otherwise with every batch email or mass distro list with your name on it, your clever little blurb will repeat fire over and over again, much to everyone's frustration. Also think twice about including a personal cell or even your Blackberry number in your reply; everyone that really needs to get a hold of you should probably have those numbers anyway.

- Auto-forward. You may want to consider changing your email options so that every email you receive also goes to your 2IC. This might sound like an efficient means of covering a lot of bases, and often it is, but for some of you cats out there, this may provide your subordinates with some interesting and revealing perspectives ("This email is to notify you that your new Dungeon Master's S&M Guide has just shipped!"). Do some analysis and weigh this decision carefully.

Getting away from work prevents burnout and allows you to recharge and apply a fresher perspective to the issues and burdens you will face upon your return. You may have 500 messages in your inbox when you get back, but they are only a 'select all' and 'delete' key away from continued bliss. If it's important, they'll write you again (this sounds cool, but it's probably a bad idea).

True Transcendence: the Selfless Team Player

AS DISCUSSED EARLIER, building teams is important. But serving as a consistently selfless team player is arguably of equal importance.

> While serving as a Company Commander, I was preparing to deploy my unit to Bosnia for an SFOR rotation. At some point, one of my Platoon Sergeants approached me with an interesting proposition. A hydration system company had contacted him and offered to set up the unit with one of their latest and greatest products, let's call it the "Shamelbak." Much to our delight, they offered to outfit the entire Company—free of charge—with their latest product.
>
> When called to attention at the Battalion's manifest formation, I thrust out my spoon–chest a little more as my Soldiers proudly snapped to, sporting their new, cool, low–profile 'Stealth' water hydration systems. We smugly stood, enjoying the subtle opus of 'oohs' and 'ahs' from the envious infantrymen assembled on our flanks, all toting the Army's prude and uninspiring standard–issue canteens.
>
> My exultant feelings were short–lived, however, when my Battalion Commander abruptly yanked me out of formation and asked me what the hell we thought we were doing with those damn things on our backs. Without pausing for a response, he sternly told me to get rid of them. Since no other Company in the Battalion had them, I guessed grade school bubble gum ROE was in effect: if you didn't bring enough for everyone, you couldn't chew it. Much to our collective chagrin, we reluctantly pulled them off and stuffed them in our duffle bags, where they remained in disuse for the entirety of the Mission Readiness

Exercise (MRE) and subsequent deployment.

The term 'team player' may at first glance seem to be a trite one, to say the least. Have little doubt, however, that it is one of the most highly sought-after attributes in any career field, especially ours where non-team players run the risk of dying. Over the past decade, board results consistently show an inclination to highlight those described as great team players within their respective organizations. As a Field Grade officer, this may very well be one of the single most important evaluation criterion with which your raters are assessing you on, if not your wingmen. And aside from the evaluations, team players save lives in the profession of arms.

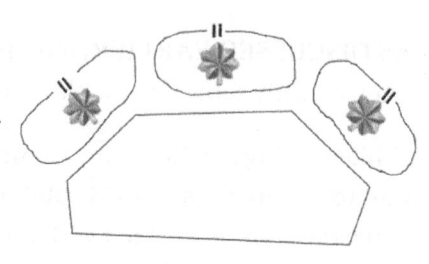

As a Company Commander, you may have huddled with your officers, NCOs, and master gunner in a darkened corner of your orderly room, conspiring on a strategy that would yield your unit that prestigious top Table VIII gunnery award. You then aggressively protected that strategy like a closely guarded state secret, visualizing the day you would hang that streamer from your guidon. As a Field Grade officer, such tactics must by fleeced from your repertoire for you to earn the deserving accolade of the consummate team player; an attribute, I would contend, that is of much greater and inestimable value to you and the broader organization as a whole.

Selfish-lessness. As a leader, frustration can occur when one of your subordinates comes up with a great idea, but keeps it on the 'down low' and refuses to share it outside of his or her own organization. Usually that reluctance stems from a selfish desire to present his or her unit in a better light, probably in an effort to enhance their own image in the eyes of subordinates, peers, and bosses. Nonetheless, the pride one unit feels over another should never stem from some garnered item, tool, tactic, technique, or lesson learned that would allow them to further their abilities in accomplishing the mission verses that of their adjacent unit—or any other. Would you really consider having those units securing your flanks to be any less prepared than your own unit? We all want our Soldiers to feel that they are part of an elite unit, above the rest, and better prepared for combat than their peers. The sense of pride instilled within your ranks should come, however, through hard, well-planned training, proudly maintained equipment, and inspirational, competent, and caring leadership, not through such selfish endeavors as petty one-upmanship. Most commanders at the operational and strategic level thoroughly understand this, and will seek to recognize and advance those that contribute their insights and gifts for the betterment of not only their own units, but also their wingman's, and that of their higher headquarters.

> *The sense of pride instilled within your ranks should come, however, through hard, well-planned training, proudly maintained equipment, and inspirational, competent, and caring leadership, not through such selfish endeavors as petty one-upmanship.*

Share your pain. When you spend several hours with your higher commander undergoing a painful Semi-Annual Training Brief (SATB), take the time to capture the commander's remarks, hot-button issues, pet peeves, and other lessons learned, and send a note to your peers giving them a heads up as to what was discussed. It will undoubtedly give those units a leg up for their own upcoming briefs, but it's not a zero-sum game. The competent execution of their SATB won't diminish your efforts. Any intuitive boss out there knows—or rather expects—that subsequent presentations within his or her own organization should get better with each iteration anyway. This is a classy and constructive habit to develop for any interaction with the boss or other higher echelons, as it just goes to enhance the overall preparation and performance of the organization. If your petulant ego won't comply with the dishing out of such free chicken for other's benefit, then simply CC the Brigade S3 or XO in the hopes that the word trickles up, cheeseball. These are the types of practices (not necessarily the 'CC' part) that truly separate the wheat from the chaff and the exceptional from the merely competent.

Imagine the collective gains to be had—both in and out of combat—by a team composed of selfless members that aspire to contribute to the greater good of Soldiers and organizations at every echelon. When you figure out and live by this truly synergistic habit yourself, and breed such attitude and habits within your own subordinates, you will have gained true transcendence.

> During my Squadron's deployment to Iraq in 2008, the insurgents in western Baghdad started to attack our patrols with an obscure, Russian-made anti-armor grenade called an RKG-3. This was a definitive shift away from the traditional Improvised Explosive Device (IED), and its use constituted a more spontaneous, face-to-face encounter with the enemy than ones we had traditionally experienced beforehand.
>
> Over the latter months of 2008 and the beginning of 2009 our Squadron was continuously humbled through successful casualty-inducing enemy attacks that, in some regard, had us demoralized and off-balance. It wasn't until we underwent some extraordinary introspection of our TTP's that we were able to reframe and provoke a tectonic shift of our own. By making some equipment and weapons modifications, in concert with some extremely modified tactics and training, we were able to completely change the paradigm, and regain the initiative decisively back into our favor. We subsequently employed a 'be the hunter, not the hunted' methodology that didn't lessen the frequency of attacks, but greatly altered their outcome.
>
> My commander quickly capitalized on our momentum, and he worked diligently to spread our successfully employed tactics among his peers

and across our Brigade's AOR. His efforts ultimately spanned the entirety of the theater, through an aggressive campaign of detailed storyboards, OPDs, internally created Mobile Training Teams (MTTs), professional journal articles, and other strategic communications. Such efforts resulted in a huge decline in casualties across the entire country where these enemy tactics had previously been employed with great effect. Not only that, but the number of insurgents captured during these attacks increased dramatically, resulting in more cracked open (and subsequently expunged) cells.

Chapter V

The Beer Math of Doctrinal Consumption

I STATED EARLY ON THAT I wouldn't dwell too much on doctrine, but to stay completely clear of it while discussing any aspect of our career field is akin to going to church without talking about God.

This chapter tackles only a few key topics, for now, and attempts to shy away from simply reprinting the inspiring and captivating prose that constitutes our dogma. Instead, it seeks to expound upon what can arguably be described as some foundational aspects of our doctrine, aspects that transcend branch, component, and in some rare cases, service. As the Army transforms yet again to cope with a dynamic operational and fiscal environment, these fundamentals must continue to stand the test of time and turbulence, and those leaders that master them will advance their units dramatically, both on the battlefield and off.

Train to Fight; Fight to Train

I DON'T WANT TO DELVE too deeply into training doctrine regurgitation here, but there are some salient points and TTPs worth mentioning that can have enormous impact on our organizations, especially as we transition back into a garrison-based Army.

During my tenure at Fort Stewart, GA, the Commanding General (CG) issued a mandate to all units within the 3rd Infantry Division that each Company should be allocated at least one week of protected Company-level training per year. Yes, that's right, one week—per year. With all of the other competing demands (Battalion, Brigade, and Division-led collective training events, gunneries, red-amber cycle obligations (details, etc.), Army Regulation 350-1 requirements, command inspections, equipment services, block leave periods, four-day weekends, etc.) it became so difficult for Company Commanders to shoehorn in their own training time on that eye chart of a training calendar, that it required the CG to enact such a radical directive. What a tragedy that such measures should have to be taken to protect training at this echelon or any other. If Company Commanders are having to fight for the table scraps of a mere handful of days per year to design and execute collective training on their own, what hope would a Platoon leader ever have in doing the same?

Imagine it. Put yourself in the boots of that lucky Company Commander and visualize a picturesque time when you were able to determine your destiny by planning and orchestrating your own training exercise. Based on your assessment, you and your leaders got to determine what key collective, individual, and leader tasks you would train, and just how you'd do it. That flexibility combined with a wide open training area, motivated troopers, vehicles, ammo and MILES gear, smoke grenades, arty simulators, parts, fuel, medics, mechanics, and hot chow—all of it, and not a Battalion weenie in sight. Your Company command post stood independent and alone, perched in a small defilade with the occasional personnel carrier booted into the side, a generator humming in the background, and hot, crappy coffee stewing in the pot. It's one of those idealistic visions that first fed the passion that drove you to enlist or commission in the first place. This is the stuff that leaders fantasize about in the wee hours of the morning... Now, snap out of it and finish off that annual online information assurance training, but don't let go of that fading daydream just yet.

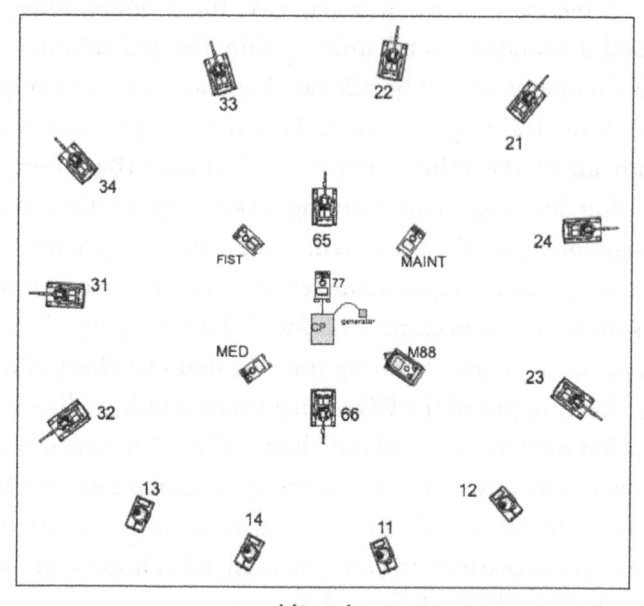

Utopia

Take an appetite suppressant. Although difficult for many staffs to achieve, the 1/3–2/3 rule (or ideally, the 1/5–4/5 rule) used during operational planning deliberately intends to empower subordinate units with the preponderance of time for both planning and preparation. Why should it be any different for the allocation of precious training time? Platoons and companies win battles and wars, so fight the urge to assimilate those rare calendar white spaces into elaborate training events at the Battalion or Brigade level. At your echelon and within the span of your responsibility, relinquish control over such time and push those infrequent opportunities down to subordinate units as often as possible (with the boss's concurrence, of course). You will be amazed at how resourceful Company leadership can be in designing rigorous, quality, Mission Essential Task-oriented training for their units when given the latitude and resources to do so.

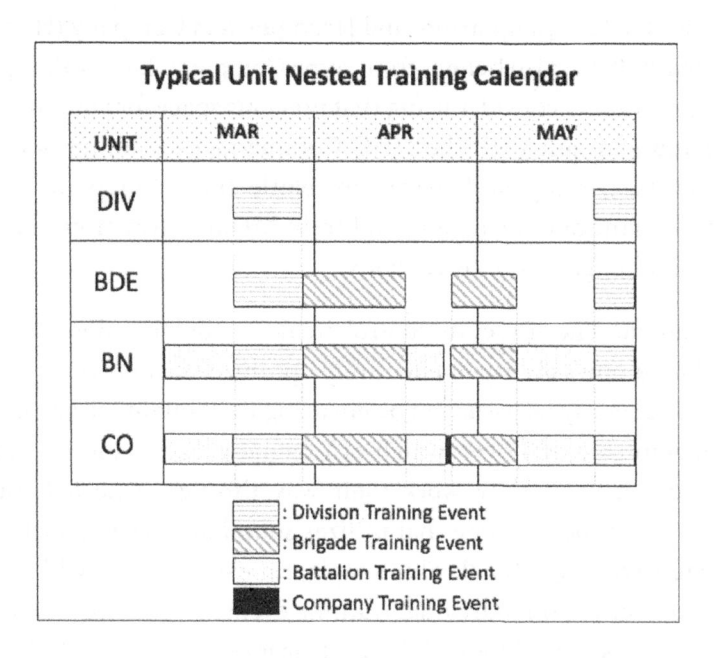

Typical Unit Nested Training Calendar

UNIT	MAR	APR	MAY
DIV			
BDE			
BN			
CO			

- : Division Training Event
- : Brigade Training Event
- : Battalion Training Event
- : Company Training Event

Events such as these will also serve as an opportunity for enormous maturation, professional growth, and experience for that Company command team, and hopefully, their Platoon leaders. Where else will they learn how to plan and execute training before they too are faced with the challenge of doing it at the Battalion Field Grade level? Those few hours spent devoted to training management instruction at CGSC will just not cut it, regardless of how great your instructor was (except Staff Group 4B in 2010 and 2011; you cats are good to go (but not really)).

Set it in stone. Ensure you plan this kind of training far enough out, perhaps by codifying it on your long-range calendar and in that vague and rarely consumed Annual or Quarterly Training Guidance, and solicit buy-in from higher by briefing it at Semi-Annual or Quarterly

Training Briefs (SA/QTBs). Be prepared to resource it at your level with training areas, logistical support, OPFOR, Observer-Coach-Trainers (OCTs), and, of course and most importantly, proper budgeting and cabbage allocation.

Nest with HHC. Headquarters and Headquarters Company (HHC) and your own staff can also benefit from some Company-level training time. You might want to consider using that precious space for your own CPX, where you can nest with the HHC Commander's training objectives, rehearse TOC setup and teardown, battle track Company training activities, manage information, and train MDMP away from the email Borg and other garrison distractions.

Power down. The insecure control freak in you that doubts a mere subordinate unit's ability to effectively optimize such precious training days can generally be assuaged by ensuring their quality usage through close mentorship and over-watch of Company Commanders. Support and guide them as they work their way through a nested training management process – one that utilizes sound approaches such as the Eight Step Training Model. This simple, timeless strategy will result in quality collective and individual training events every time. Make these training events special topic slides for Brigade-level Quarterly Training Briefs, where Company Commanders can have an opportunity to prove their metal (METL?) in front of their senior rater, who might just be inspired enough to visit such a well-crafted training extravaganza.

The Sacred Six. A sound unit training management process is of ludicrous importance, as is embracing and enforcing a six-week lock-in for Company level training schedules. Ensure essential resources for Company training events are available no later than seven or eight weeks out. Don't put the cart in front of the horse on this; if you try to resource your training after you lock it in on your training schedule, you'll jeopardize its execution (e.g., you can't go to the range if you don't even own the range or have the ammunition allocated). It is also worth noting that resources and personnel requested inside of six weeks may very well disrupt another unit's training plan (e.g., booking your Company medics for a training event that is four weeks out is well within the HHC Commander's locked-in planning range). Use Battalion training meetings to have companies brief their T-8 (broad concept) and T-7 (scheme of maneuver) training week schedules. Solicit additional commander's guidance, resource requests, and tentative approval; then provide refined T-6 week schedules for signature (usually through DTMS). Changes to the training schedule within those six weeks should be minimal at best, but, as the saying goes, feces happens.

Defend in depth. In order to protect these 'training contracts' between a Company and its higher headquarters, abide by a policy of escalating commander's approval for changes, at least within your span of influence. For example, Battalion Commanders might approve changes to week T-5; Brigade Commanders approve required changes to T-4 and 3, and perhaps echelons above that for changes within two weeks (it was not uncommon in our Brigade to submit a memo to a General officer in order to cancel or modify training for the following week—a distinctly painful endeavor, I can assure you).

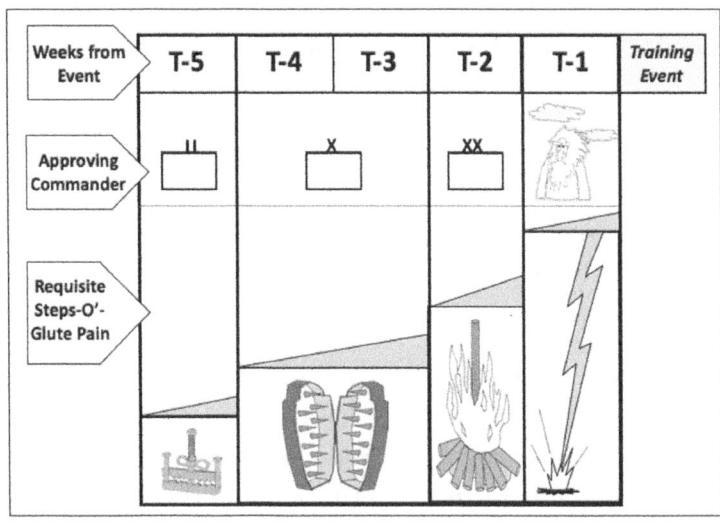

Weeks from Event	T-5	T-4	T-3	T-2	T-1	Training Event
Approving Commander	II		X		XX	
Requisite Steps-O'-Glute Pain						

Change-to-Company-Training Consequence Spectrum

The intent of such a process is not to heap additional work onto the shoulders of Company Commanders, but rather to protect them from haphazard and disruptive changes that most often are shoved down their throats from Battalion or Brigade. The authority that drives the change should bear the onus of initiating the approval request memo. If Battalion determines that it must harvest another Platoon from Alpha Company to support a detail during T-3, the Battalion S3 shop must draft a memo signed by the Battalion Commander to the approving authority (Brigade Commander) in order to allow that change to A Co's training schedule. The shame in having to explain their inability to plan accordingly to one's higher headquarters, coupled with the predictable backlash, will greatly discourage such incidents in the future, and further reinforce the sanctity of the Sacred Six.

Infraction traction. In addition to echeloned approval for changes, Companies and Battalions should capture deletions or training modifications in their Quarterly Training Briefs through the use of the ominous 'training planned but not conducted' slide. This dime-dropper medium tends to raise both eyebrows and hackles from senior leaders—a good thing for those who did their due diligence through sound training management, and a bad thing for those who failed to conduct the requisite analysis and long/short-range & near-term planning.

Master UTM. Along with our training doctrine, consult with the handy pub: A Leader's Guide to Unit Training Management, for further details on making the most out of your training calendar. This well-crafted manual can be found on the Army Training Network (ATN), and is somewhat more concise and intuitive than the associated ADRPs.

The imperative. Running effective, nested training meetings at Company, Battalion, and Brigade level is vital to ensuring quality, mission-focused, and resourced training. The training meeting is probably the single most important meeting of the week at both the Battalion and Company level. The art and science of running a concise, focused, and effective training meeting has, in great measure, been lost upon us over the past decade, especially at the Company level. The trusty and crusty TC 25–30 was a good companion, and thankfully most of it can still be found in another great reference on ATN: A Leader's Guide to Company Training Meetings. This succinct publication will help your Company Commanders construct quality meetings within their organizations (take note of the videos and links for further information).

Resource it. Employ a weekly Battalion training resource meeting. This handy training meeting time saver is chaired by the Battalion XO, and includes Battalion operations and logistics reps that are empowered to make decisions about resources (master gunner, range/ammunition manager, maintenance officer, other key forward support Company personnel (XO & support Platoon leader), medic OIC/NCOIC, etc.). This is a great forum for training event sausage making, where Company First Sergeants and XOs can coordinate directly with higher's providers and deconflict support without commanders at both echelons breathing down their throats. Keep this meeting scheduled toward the end of the week following both Battalion and Company training meetings. This will allow all attendees to bring in their requests based on the latest approved broad concepts (T–8) and schemes of maneuver (T–7) that were just briefed to the Battalion Commander only a couple of days prior. It also gives companies another opportunity to synch their just–signed T–6 Company training schedule requirements. Create and utilize a training event support worksheet that can be used as a checklist in codifying all forms of training sustenance. Carry these sheets through each week's meeting until all forms of support are answered for each event. Doing this kind of legwork weekly will ensure both Company and Battalion training meetings stay focused and brief.

Train the training meeting. As a Battalion Field Grade, don't be afraid to crash a Company training meeting on occasion. This is a primo opportunity for you and other members of the Big Five to observe how your training priorities and policies trickle down to and affect Companies and Platoons. But more importantly, you get to serve as an ad hoc OC/T, and, following the event, you have the opportunity to meet with the Company Commander for a quick AAR and some involuntary mentorship. Not only do these visits serve as azimuth checks for subordinate units, but, based on acute Battalion leadership interest, it's also a shot across the bow for Company Commanders and First Sergeants that will convey the importance of scheduling and actually conducting quality training meetings.

> *As a Battalion Field Grade, don't be afraid to crash a Company training meeting on occasion.*

Recently our Commanding General shared with us that he pulled ten Company training schedules from across the installation. He identified their posted weekly training meetings and visited every one—unannounced. Only two of the ten were executed in accordance with their schedules, or at all. Those that did occur simply spent their time clumsily reviewing the next day's training for a few minutes. You can probably imagine what happened to the other eight Company Commanders, especially in light of the hate mail and fecal storms that ensued at the Battalion and Brigade levels. Such embarrassment can result in career-altering propagations spanning multiple echelons.

Don't let your commanders become members of that 'Lucky Eight' club. As training once again becomes the most important thing we do as a garrison-based Army, don't be surprised when Brigade and Division Commanders pursue a similar policy of Company and Battalion training meeting drive-bys in your unit.

A parting thought on training...

In the late '80s, I enlisted and served as an Infantryman after high school. During that time, under the bennies of unpredictable and sometimes abysmal funding, I gained a true appreciation of what it meant to train in an Army that was not considered to be a fiscal priority in anything but lip service. With perfect clarity, I can still recall weeklong field exercises (we called them 'field problems' for reasons I could never figure out) where I was given a single box of 20 x 5.56mm blank cartridges, and told to 'make the most of them.' We tended to blow through that ammo in the first minute of our first 'enemy' contact, and we spent the next six and a half days shooting at the opposing force (OPFOR) with plentiful foul language and adamant 'bang–bangs, you're dead.' The M–60 machine gunner went horse by the end of first engagement, and would simply curse having to lug that empty, silent pig around for the rest of the week. I felt like I was playing in the woods as a third grader all over again. We would literally break out into juvenile "I shot you first,' "no I shot you first" yelling matches in the middle of an attack. A good portion of my squad never even pulled the trigger, but quietly buried their few measly blanks in the dirt to keep from cleaning the invasive carbon off of their weapons. Perhaps 'problem' was an effective term for such an event after all.

Well, sports fans, as much as I hate to say it, there's a good chance those days are a comin' back. In the vicious cycle of wartime appreciation and political 'peacetime' neglect, the storm clouds of military budget cuts gather once more in greater frugal density, and their inevitable implications will instigate a tectonic shift few young Army leaders and Soldiers could possibly imagine (alliteration aside). When these dismal days inevitably descend, we will be granted no quarter in continuing to bear the burden of our mission and that of national defense. "Do more with less" will be the mantra.

Keep in mind that when the call comes—and it inevitably will—you have to cross the LD regardless of how much money you weren't given during training.

It will take enormous innovation and creative solutions at every echelon to make the most out of every training dollar to ensure our units are capable of accomplishing their mission. For many of you, this will be your crucible task: "how do I get and keep this Battalion trained enough to execute its wartime mission under these dire circumstances?" Your answer to the unit should never be "I can't" despite all the ruckus and fervor you raise to the contrary to higher while trying to candidly convey your plight. Don't compromise on your USR and other assessments. Call it like it is, but work tirelessly to exploit every potential training opportunity, funded or otherwise, to get your units combat ready. Always keep in mind that when the call comes—and it inevitably will— you have to cross the LD regardless of how much money you weren't given during training.

Think Outside the Maneuver Box, but only after You've Learned to Think Inside it

I BRIEFLY MADE MENTION of this in my introduction, but I think it is of such great importance that it's worth mentioning again: as a Field Grade officer, you must establish the attempted mastery of doctrine at your level and position as a permanent line of effort in your professional campaign plan. This fundamental duty of warfighters—especially Field Grades—is inherently a personal responsibility, and must be continued well beyond the hallowed halls of Professional Military Education (PME) such as CGSC.

As a Field Grade officer, there is a general expectation from your subordinates, Company Commanders, and assuredly the Battalion and Brigade Commanders, that you are the Doctrinal Dumbledore of the organization, where at a moment's notice, you should be able to regale an awe–inspired crowd with a definitive statement on the subtle differences between the tactical tasks of 'seize' and 'secure'.

As we've discussed in multiple chapters of this book, there are plenty of opportunities to show your ass as a 'Bush League' amateur, but few will undoubtedly detract from your subordinates' confidence in you more than chronically misusing, misapplying, or misrepresenting doctrine in any forum. On a day-to-day basis, there are probably several shabby facets of your Field Gradeship that most subordinates are willing to forgive, but your unsound or botched employment of doctrine in an operational context is not one of them. What does that mean? Well, it's certainly no feat of intellectual doctrinal capacity to real off the (19 and counting) steps of Mission Analysis, but within your respective career fields, staying current on and well-versed in doctrine will provide you with some of the requisite tools (the science of our profession) necessary to successfully and artfully fight your units inside the box and, ultimately, beyond it.

Master the lingo. There are only a few letters difference between 'decision point' and 'decisive point', but you can bet they are worlds apart in meaning and context. How can you adequately describe your scheme of maneuver to anyone, regardless of branch, if you don't speak or understand the ever-evolving language? This integral comprehension of the lexicon starts with our internalization of military operational terms, graphics, and symbols (note that I didn't site the pub nomenclature, as by the time you read this, it may well have been renamed again). As a Field Grade officer, there is a general expectation from your subordinates, Company Commanders, and assuredly the Battalion and Brigade Commanders, that you are the Doctrinal Dumbledore of the organization, where at a moment's notice, you should be able to regale an awe-inspired crowd with a definitive statement on the subtle differences between the tactical tasks of 'seize' and 'secure'.

Know what to know and where to go to know. There are two broad categories of doctrinal knowledge: 'know it,' and 'know where to get it.' The former is stuff you should know off of the top of your head, regardless of where you are and the latter, stuff you can afford to look up in a reference. What determines which category a piece of doctrine belongs to? I'll contend that your mission statement, mission essential task list (METL), and the most demanding parameter—time—are probably the biggest screening criteria.

Let's evaluate the relatively simple matter of understanding the difference between the terms 'attack' and 'attack by fire.' When conducting some water-cooler doctrinal debate, one might have the luxury of executing a quick internet search to bring up the definitions of both terms, and subsequently make the call as to what's appropriate for the contentious argument at hand. I would submit, however, that there are plenty of instances where time is not an indulgence you can afford, and your inability to discern the categorical differences between those terms may have immediate and dramatic consequences for your unit.

So, when you're sitting in the TOC sucking down another sleeve of thin mints, and that contact report comes in hot over the radio, do you have the time to research what type of mission or tactical task should be issued out to the unit? Will the Company Commander wait patiently on the radio for you to generate that FRAGO as you flip pages in the pub or search the internet for an answer? Will the unit even know what you mean when you say 'attack that building,' or will they remain silent, waiting for you to specify a tactical task and purpose in conjunction with that type of operation? Will they continue to close with the enemy upon given the task to 'attack by fire?' Would you? If you suspect you would answer any of these questions in a dubious fashion, then perhaps you've uncovered some 'know it' doctrinal material.

The 'know where to get it' group is broad indeed, and includes more conceptual or foundational doctrine, problem solving processes, technical data, unit training management procedures, etc. In no way, however, does this category exempt the warfighter from possessing a general familiarization with such dogma, as well as the knowledge of what pub one should go to for greater, more in–depth enlightenment.

Stay current. This can be a greater challenge than it would seem at first glance, especially when our careers demand that we periodically dive in and out of tactical and operational units while seeking other broadening assignments. You might be surprised at how your branch-specific or Army doctrine will reinvent itself in a cycle of improvement that was once unheard of. Today's turbulent operational environments evolve at a rate that legacy doctrinal production could never hope to keep up with. Now it's not unusual to have a two–year old publication on your shelf that is vastly out of date, along with a good portion of its terms and processes.   Things were so much easier when most of our doctrine was designed to stop and destroy the rather predictable and unimaginative Soviet threat (CRP—FSE—AGMB—VIC-TOR-EE!).

Rehearse Well

REHEARSALS ARE NECESSARY in order to soundly execute anything from ceremonies to combat operations. For the purposes of this section, I'll dwell specifically on rehearsals in support of tactical operations, arguably one of its most important applications. If these can be mastered, then all other event rehearsal types should be that much easier. The 1998 Center for Army Lessons Learned (CALL) handout on rehearsals was a breakthrough in codifying everything we knew about conducting them into a single, concise, well-crafted document. Thankfully, it continues to live on today in a few pubs, and can now be found almost in its original entirety in the latest version of FM 6-0. Don't worry; I'm not going to reprint it here (but do go and read it). I would, however, like to expound on a couple of points, and identify some potential pitfalls in rehearsal preparation and execution.

Salmonella chicken. When your rehearsal turns into a wargame, your half-baked COA analysis hens have come home to roost. There's a lot of talk about using the rehearsal to synchronize the plan, even in the field manual. I certainly agree there is room for tweaks here and there, especially as this will be first time that everyone with a vested interest in the plan can finally visualize it in all its glory, oftentimes depicted in three dimensions upon a complex and roomy sand table. I also offer, however, that the time to 'synch' the plan is around the wargame table during COA analysis, and not in the even more time constrained and daylight-consuming environment of the rehearsal. Every minute spent debating the plan around the terrain board is one less minute that those captive commanders get to spend readying their own units for combat. So, in some regards, it's hard to talk about successful rehearsals without talking about successful wargaming—the precursor to any concise, quality rehearsal. Let's briefly discuss.

War-game, schmore-game. Ever met anyone who liked this step of the Military Decision Making Process (MDMP)? Outside of the sadistic Small Group Instructor (SGI), there are few who would relish the thought of six to eight hours of sweaty bickering, bad coffee, and listless, irritable staff choagies gathered around a cartoonish mapboard covered with unit icons that seem to creep lethargically toward the objective with the hustle of a hamstrung sloth. But here is where the real synchronization happens (or should happen). In order for this to occur, however, staff members must look at the plan through the eyes of those absent yet wholly vested unit commanders, and work the details out on that painful synch matrix through all Warfighting Functions and every critical phase of the operation. If time allows, this effort should also include branches and sequels tied in with commander's decision points. Depending on the experience of the staff and the effectiveness of unit SOPs, you might have to work over those non-critical phases of the operation as well (e.g., moving from the TAAs into attack positions, etc.).

Pay your penance up front, and make your staff do its due diligence around that frustrating and often exhausting wargame table to avoid turning your rehearsal into the eighth step of the planning process.

The synch matrix, if effectively completed in COA analysis and combined with sound unit SOPs, will transform into the operational execution matrix for those fighting units soon enough. It's this key product that commanders at all levels will use in the conduct of the rehearsal, along with their own derived planning products. Its lack of coherency and completeness will bear rotten fruit at the rehearsal while you're trying to orchestrate the attack with all those commanders and coordinating staff members poised on the terrain board, icons in hand, curious as to where they should bound or who's got the next move. Pay your penance up front, and make your staff do its due diligence around that frustrating and often exhausting wargame table, to avoid turning your rehearsal into the eighth step of the planning process.

OPSEC it. The ease in which an enemy can employ drones or 'civilians' to gather intelligence on our preparation efforts makes it much more difficult to protect the vital information our rehearsals contain. This is especially true of large, outdoor sand tables accompanied by maps, overlays, microphones, and booming loudspeakers, as well as full-dress rehearsals, where vehicles and personnel 'act out' the plan in broad and sometimes all-too-evident fashion. An overhead snapshot of our operational graphics depicted on replicated terrain with engineer tape, spray paint, and yarn can serve as an invaluable piece of intelligence for the bad guy. Ensure you consider and plan against every possible traditional and modern means that the enemy may use to exploit your rehearsal in order to gather information on your upcoming operation.

Know the Adapting Enemy better than the S2 (Unless You're the S2)

WE'VE ALL HEARD that Intelligence Preparation of the Battlefield (IPB) is 'everyone's responsibility,' but we often pay it too much lip service, and look squarely at the S2 to bail us out during planning. That's an unfortunate crime, whether you're serving on staff or in a command capacity. Regardless of branch, echelon, or staff section, this practice should be something you have internalized from top to bottom. Then, when you have the process down, become your own S2. Defining the problem oftentimes starts with defining the enemy first, regardless of echelon. Savvy tacticians fight savvy with both blue and red hats on.

Spread the wealth. Regardless of what your branch is, you can't punt analysis of the enemy to the S2 in its entirety. Who's the best officer on staff to discuss enemy air defense capabilities—the S2, ADA officer, or the staff aviator? I would submit it should be the last. On a daily basis, it's the aviator's life that's at risk, so the plan will depend on how he or she contends with the enemy's employment of these systems against him or her. In the same regard, the staff's ADA officer should be the resident expert on how the enemy will most likely employ its air assets against us.

Be the ball. Certainly Platoon Leaders, Company Commanders, and even many Battalion staffs won't have the luxury of a Military Intelligence officer as an S2 to assist with their respective IPB. Ideally we will receive some analysis from our higher headquarters, but the obligation to continue the process of drilling down to the appropriate echelon and to provide the proper level of resolution on that echelon will largely fall on unit commanders, especially at the Company and Platoon levels. As a result, small unit commanders will have to be experts on every form of contact the enemy can bring to bear, and be able to overlay that capability on terrain and weather relevant to the operation. When you start to do the math on this, it seems a bit daunting. Have no doubts that it is. Your average maneuver leader, for example, will have to be the masters of the following:

- how enemy armor will be employed
- effects and ranges of their weapons systems

- when enemy reserves will commit (triggers)
- when forces will reposition or withdraw (triggers)
- how infantry will be employed and what anti-tank capabilities they possess
- where indirect assets at every echelon will be placed to affect the operation
- where IEDs and VBIEDs might be positioned to disrupt formations
- how other manmade and natural obstacles will be employed to shape engagement areas
- when and where enemy air assets and missiles will be committed to support the fight
- when and if jamming will occur, and what systems of ours will be affected (UHF, GPS, etc.)
- will the enemy utilize chemicals, and if so, when, where, and what type
- how civilians might be by exploited by the enemy to disrupt the operation

Sell it. Our obligations don't end with the acquisition of this knowledge. One must master the art of conveying that information to units, commanders, and staffs in a fashion where it is understood completely, just like their own execution paragraph. This is a necessary, art-heavy task that requires leaders to 'sell' their interpretation of the operational environment. Over my years instructing at the Armor Captain's Career Course and CGSC, I was privileged (relative term) to witness countless IPB briefs and paragraph 1 analyses from truly brilliant officers and staffs. The biggest challenge for many to overcome, despite their sound interpretations of the enemy and terrain, was an inability to create the requisite products and orchestrate an intuitive approach to conveying that critical information to their respective wares. Spend the time making visual products that easily communicate how you expect the enemy will fight you. In addition to SITEMPS, sketches, and PowerPoint slides, these renderings may be incorporated within the sand table, displayed on small, portable dry-erase boards, or chalked on the side of an M113 (employ some OPSEC).

Spend the time making visual products that easily communicate how you expect the enemy will fight you.

Picture this... Develop a clear method for visualizing just how and when the enemy will attempt to mass forms of contact on your unit. You might consider scripting the ECOA relative to time and space ("as we approach LD, enemy reconnaissance elements will gain visual contact around the 23 easting or perhaps as far back as our attack position through the use of unmanned aerial systems; our lead echelon may start to encounter some sporadic and possibly inaccurate effects of long range artillery shortly thereafter, while the enemy simultaneously triggers IEDs along the MSR in an effort to disrupt or delay our trains"), or perhaps by discussing, by each form of contact, how the enemy will fight us through the depth of the operation.

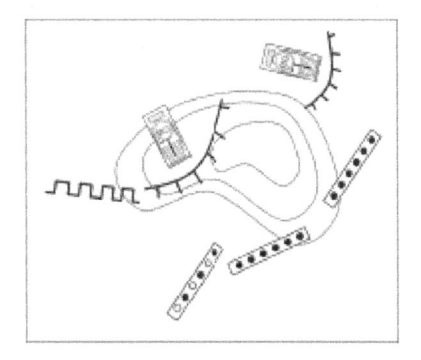

Own Intent

THERE'S A GOOD CHANCE many of you out there will assume command of a Battalion someday (especially after reading and abiding by this fine work). When that time comes, I most strenuously encourage you to implement some or all of the following TTPs.

Don't pawn it. When you were a Company Commander, who wrote your commander's intent paragraph? Your XO? Perhaps your First Sergeant? Yeah, probably not. While constructing your Company operations orders, you'd be damned before you dished off your intent paragraph on some other moe-frackie. Why should it be any different when you take command of a Battalion or any other unit? The commander's involvement in MDMP is traditionally inversely proportional to the staff's experience and competence. But regardless of how much faith you have in your stellar staff, take the time to generate your own intent paragraph. After all, it's called commander's intent for a reason. Ok, I'll confess: I'm a couple of beers into it as I write this paragraph, and reflecting upon all of the orders briefs I attended where the commander was content to sit on his can while the staff developed and briefed his intent is making me a bit (more) red in the face. Flagellate yourselves, gentlemen; you know who you are.

> *Regardless of how much faith you have in your stellar staff, take the time to generate your own intent paragraph.*

Speak it. Here's some more crazy talk: in addition to constructing the paragraph yourself, you actually choose to stand up at the mapboard and brief it during the OPORD issue. Is it not awkward to hear the S3 brief your intent while you look on sheepishly from a defilade position seated in the front row? If you want to run away from the tedium and pain of the MDMP to conduct some quality 'battlefield circulation,' that's fine, but do so only after you have sat down and put some thought into this oh-so-important paragraph. At the end of the day, if the members of your unit forget everything else, they can still operate towards mission accomplishment with the impactful, lingering memory of your emphatically delivered, concise statement on why we are getting out of bed in the morning and what conditions will define success for us all.

Groupthink it. Sure, make the S3 and XO generate proposed intent paragraphs as well. How else will he or she get trained in the process? Then, take a few minutes to gather your Field Grades and conduct a little compare and contrast with your own product; consolidate and edit accordingly, then get it into the order (or hopefully, the WARNO).

Differentiate it. Oftentimes we use the terms 'intent' and 'guidance' interchangeably. Although interrelated, their meanings vary quite a bit. There are plenty of touch points where we seek the commander's guidance through the execution of our planning process. This can range from parameters on the planning process itself (e.g., "develop and wargame two courses of action; one for the enemy's most likely COA, and one for his most dangerous") to mission-specific guidance following the mission analysis brief such as the employment of Field Artillery assets and other warfighting functions in support of the maneuver plan, etc. This guidance can include, but should not be confused with, the commander's intent for the operation.

Lessen it. The intent paragraph is yet another opportunity for you to take an appetite suppressant. Again, you're not being graded on volume, and you're not briefing the entirety of the scheme of maneuver or execution matrix, so don't feel like you need to write a dissertation while constructing it. It's worth noting that our (Army) doctrine has become more open-minded on how an intent paragraph should be composed— at least for now. Like current joint doctrine, Army field manuals used to recommend a 'purpose, key tasks, and endstate' approach. As of late, however, Army doctrine has become a little less prescriptive in mandating (or rather specifying) the inclusion of key tasks in the intent paragraph. This was probably as a result of a disturbing evolutionary trend that distorted the concise spirit of the intent paragraph by swelling it into a tome of sorts, which merely duplicated other parts of the 'concept of the operation' paragraph. I can recall Division and often Brigade Commander's intents that were a full page or two in length, spelling out every (key?) tactical task and purpose that must occur from uncoiling the tactical assembly area to consolidating on the objective. I'm sure they were all important to the operation, but most—if not all—could be found in more detail and in better synchronicity through more appropriate paragraphs like 'scheme of maneuver,' etc.

The purpose of purpose. At this point, 'purpose' is not an optional part of the 'intent' paragraph. That being the case, consider the following TTPs in the generation of your 'purpose' sub-paragraph:

- **Task or purpose.** Don't pepper it with tasks or types of operations, e.g., "the purpose of this operation is to destroy the enemy company on Objective Dog." That's a tactical task, not a purpose. There's a reason why that Company needs to be destroyed, and it's probably explained in the second half of the mission statement already e.g. "...in order to protect the western flank of TF 1-64, the Brigade's main effort." It can be restated here, or you can address a broader purpose, if there is one.

- **Broader purposes.** There may or may not be an evident broader purpose for the operation. The one spelled out in the mission statement can often suffice. However, there are times when the appreciated brevity of the mission statement might leave out a broader explanation as to why the operation is being conducted. An example of a broader purpose in this scenario might be "we will protect the flank of the main effort by causing the committal of the enemy's reserve towards Objective Dog." Perhaps the destruction of a defending enemy Company is the perceived trigger for the higher enemy commander to commit his reserve. In sending it to OBJ DOG, it potentially is no longer in a position to affect the attack of TF 1-64, thus we have succeeded in protecting its flank.

Apply a 'key task' litmus test. As mentioned earlier, commanders often feel they must generate key tasks by executing one of the following suspect habits within their intent paragraphs: (a) regurgitate the mission statement or elements of the scheme of maneuver, e.g. "we must destroy the enemy Company on OBJ DOG," or (b) simply state the obvious, e.g. "we must successfully uncoil from the assembly areas." Both serve no purpose in the intent paragraph besides to weigh down your product with irrelevant or redundant verbiage. In the first example, the tactical task of 'destroy the enemy MIC on OBJ DOG' has most likely been stated clearly in the mission paragraph. The second example is inherent in the operation, regardless of what tactical tasks we plan on prosecuting on the battlefield.

One of the questions you should ask yourself in determining whether or not to list or describe key tasks in your intent is 'what must my unit do to achieve this purpose, regardless of the scheme of maneuver or changing conditions on the battlefield?' Over the course of an operation, tactical tasks may come and go, but the purpose will likely remain the same. So, don't fall into the trap of listing key tasks that aren't reverse engineered from the purpose of the operation. You may also want to consider key tasks that must be successfully accomplished in order to achieve a decisive point e.g. "we must suppress the enemy on the far side of the obstacle in order to protect the breaching effort." Note that key tasks need not be certified tactical tasks either. I would argue that those that are will probably be captured in the scheme of maneuver or tasks to subordinate units already (destroy, seize, secure, etc.). So, given these vaguely described and potentially contradictory constraints, it becomes a much greater challenge to generate necessary key tasks for inclusion within the intent paragraph. If you do the critical thinking and come up with nothing, then great, don't sweat it—move on to deriving the conditions that define success at endstate.

Own it. Intent has always been one of the most important aspects of an operations order at any echelon. Through the spirit of and emphasis on mission command and mission-type orders, it's even more so now. As the commander, you stand as the most experienced guy or girl in the room (or tent); so let it shine through in your own thoughtfully constructed and personally issued paragraph.

Chapter VI

Live to Fight another Day

WITH THE EXCEPTION of some parting thoughts, this chapter focuses almost entirely on different aspects of our career field, and potentially on how to navigate through some craggy straights. In the greater campaign plan that is your life and profession, maintaining a line of effort oriented on sustaining and shaping your career (a k a the "deep fight") is essential to ensure you can keep up the struggle well beyond the setting of the day's sun.

Don't Estrange Yourself From Your Senior Rater

I WAS AT THE END of my rating period as a Cavalry Squadron XO, and today I would endure my final OER counseling with the Brigade Commander, my senior rater. It was to be his second evaluation on me, and, as a Major in a Key and Developmental (KD) position, it was an important one. As he invited me to sit down, it occurred to me that I had probably exchanged only a few dozen words with this man over the almost two years we had served together in the unit, and I was pretty sure he couldn't point me out in a line-up. Apparently, he felt the same way, as he started the session quite bluntly by stating, "you know, Dave, you're the XO of the best unit in this Brigade, and I have no doubt much of that success can be attributed to you...but I don't know you from Adam." With some effort, I remained stoic and thought sardonically, "yeah, but whose fault is that?"

It would take me a couple of years before it would dawn on me that it really was my fault, in a great many regards. In retrospect I had always maintained this idealistic notion that it was the leader's responsibility to know his or her subordinates, not the other way around (I know, crazy talk). Fuelled by self-righteousness and perhaps too much pride, I certainly didn't help myself by stubbornly standing back in the shadows waiting for it to change, no matter how hard or well I worked behind the scenes. In further reflection, it was akin to an occasional and ignorant philosophy I had exhibited in college when taking a class from a poor professor: "I'll punish you for your ineptitude by doing crappy in your crappy class." The end result in both instances was pretty evident: I merely did myself a continued disservice and it would be one that would have implications on my future.

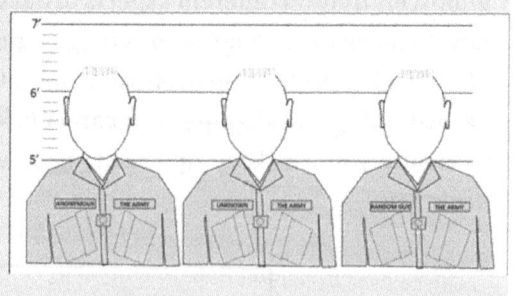

In no way is this meant to be a self-aggrandizing or self-promoting tip, and it certainly doesn't imply that you need to be a cheeseball. As Field Grades—especially deployed Field Grades—you may find yourself further and further removed from your senior rater (stationed on an isolated JSS, a distant Afghan province, etc.). You may also find yourself in an entirely different country or post than your senior rater. This is especially true for low-density MOSs. In many regards this is a good thing. It does, however, prevent a person who contributes enormously to your potential professional future from knowing who you are. As a Platoon Leader or Company Commander, your reputation was tied directly to the success (or failure) of the unit. If your Company did well (e.g., gunnery, CIPs, shucking IED emplacers, clearing routes, seizing objectives, etc.), you were often recognized through association by your senior rater as a successful leader. As a Battalion Field Grade and member of a staff, these highlights and quantifiable achievements are less linked to you as the S3 or XO, and more associated with the Battalion as a whole. Few people would look at a Company that just successfully seized an objective and think, "Wow, I bet that Company has a great XO working behind the scenes." In the same regard, few Brigade Commanders will think, "Wow that Battalion did really well during our Command and Staff meeting; I bet their XO is the bomb!"

Of course, a good boss/rater will show off the efforts of deserving Majors and staff leaders that contributed to the unit's success, but in many ways you are responsible for adding depth and dimension to a vague name atop an OER on the Brigade Commander's desk once a year. So, how do you do that without being cheesy? Simply enough, through sound, basic Army etiquette. Here are some tips:

- Don't deliberately avoid your senior rater, i.e. don't head for the port-a-John when his PSD drives through the gate of your JSS. Make it a point to greet him; serve as a guide and show off your unit and its Soldiers to him.
- At unit social functions, don't isolate yourself and expect your

senior rater to come to you. Be proactive: approach, greet, be courteous and succinct, and then get out of the way.

- Provide a good OER support form that adds depth to your personality by addressing both professional and personal goals such as family or hobby/sports achievements (i.e., coach Little League, run a half-marathon, spend time with daughter, read This Kind of War, etc.).

- As discussed in some length earlier, one of the best ways to be recognized for what you are is to be a team player beyond the span of your own Battalion/unit. Everyone will know who supports the team and who doesn't. All will cherish the selfless staff leader or Battalion Field Grade that aggressively shares ideas, lessons learned, and information throughout—and for the benefit of—the entire Brigade/unit.

Don't be a stranger; work hard for your unit's and the Brigade's success. Your Soldiers will benefit, and you will or should be appropriately recognized when the time comes.

Understand the Army's Career Fields and Functional Areas

AS A JUNIOR CAPTAIN serving in the Brigade S3 shop, I was faced one day with the suspense of completing my functional area preference sheet. Not really knowing one area from the next, nor really caring, as I was always going to be an operations guy, I yelled down the hall to one of my peers, "John, what did you put for your functional area request?" "Forty-nine," he yelled back. "What's that?" I asked. "Operations Research & Systems Analysis," came the reply. Ah! I thought, that sounds clever, plus it has the word 'operations' in it, so it's got to be cool. I quickly scribbled '49' as my second choice, right under 'operations,' then a smattering of other random numbers in the subsequent boxes, all of which I had no idea what they meant beyond their vague titles. Do I really need to know what these areas are anyway? I'll be damned if I ever leave operations.

I would come to regret my haste several years later when one day I picked up the Army Times to look up my name on the Major's promotion list. 'Everyone likes to see their name in the paper,' I happily thought, as I drew my finger down the long list until I found Dunphy, David. I was somewhat befuddled—if not alarmed—however, to find 'FA49' next to my name on the list. A quick call to Armor Branch would indeed confirm that I had been drafted, quite involuntarily, into the ranks of the big-brained ORSA community.

Several months later, while trying to decipher what some eccentric, coke-bottle glasses wearing college professor was spouting on about in an advanced, graduate statistics class, I realized I had indeed made a huge mistake by ignorantly putting FA49 down on that damn preference sheet. Yes, I had been an engineer undergrad, but I was the guy that bought all the pizzas in my design groups while my savvier classmates worked the hard numbers. I couldn't tell the difference between derivatives and dirigibles, and I was in way over my head as an FA49, all because I didn't take the time to research the Army's Career Functional Areas a few years earlier. It goes without saying that my time and humiliation as an ORSA were short lived, and thankfully I was quickly traded back to Armor Branch, probably for a six-pack of Meister Brau, or something of lesser value.

Conduct Recon. Before attending the Captain's Career or Advance Course (or whatever the hell we're calling it these days), do the necessary research on the Army's other career fields and Functional Areas. Some may serve as only temporary path changes with a re-plug back into the Operations Matrix, others you may embark upon for the rest of your time in the Army.

Fork(s) in the road. There is life beyond Operations and Operations Support (base branches). Know when your professional timeline can incorporate the 'Blue Pill' decision points of career field or functional area designation, and thoroughly understand what you may be getting into. You might find yourself one day burned out at the prospect of another decade in your base branch. There are plenty of skill sets, broadening experiences, and educational opportunities available that might appeal to your own abilities and desires, and allow you to serve in capacities well beyond the mainstream.

Prepare for the draft. Don't rule out the possibility that you may find yourself drafted into a Functional Area (see anecdote above). Do the math ahead of time, and put some analysis into the completion of your career field preference documents or equivalent products when the time comes.

Regardless of your own career path decisions, as a Captain and above you need to understand those other career fields and be able to discuss them with your subordinates during professional growth and timeline counseling.

Stuck in a Purple Haze: Notes from the Joint World

AT SOME POINT IN YOUR CAREER, you may find yourself preparing for a joint assignment or service as a member of a Joint Task Force. Before donning that dubious purple garb, consider the following random observations.

> Shortly after assuming the Joint Task Force Guantanamo operations officer role in 2011, I remained perplexed at the absence of a daily tasker or Frago within this unique command that would adequately manage and distribute information, enforce compliance, and provide some predictability for our subordinate units throughout the JTF. I approached my boss, the JTF Chief of Staff and a Navy Surface Warfare (SWO) Captain (O6), and laid out the case for a centralized document issued daily from the operations directorate that would include key

information, taskings, and other operational requirements.

The modus operandi before then had typically been an email from an interested, random staff section sent through a mass JTF distribution list that spanned all echelons; hardly an efficient system, but one that, arguably, was rather simple. When I explained the Daily Frago concept, my boss remained suspicious, if not downright skeptical. "Why should we consolidate all of this information through one shop, add yet another step, and distribute it through yet another document?" he asked. "If we have to distribute information on a ship, we grab the 1MC (microphone) and put it out to the entire crew; makes things nice and simple. Through email, the staff does pretty much the same thing here." This concept, however logical to him, simply contradicted 20 years of my own experience, and flew in the face of all things right—at least in my Army mind.

After two more pitches, including a comprehensive PowerPoint presentation (with animation, nonetheless) on the benefits of a Daily Frago, I was told definitively to pound sand; there would be no such bureaucratic product while he was the Chief of Staff. This clash of cultures was indicative of daily life within the Joint Task Force, and it really forced me to take off my Service horse blinders. It kept things interesting though, and that wasn't necessarily a bad thing.

Extra credit. Depending on your branch or functional area, joint credit may be required before promotion to certain grades can be achieved (typically, before GO/FO grades). Not all 'joint assignments' are eligible for joint credit, so be cautious as you manage your timeline. Accredited joint jobs can be found in large part on the JDAL, or Joint Duty Assignment List. Work with your mentors and career managers to figure out when the time is right for you to pursue this kind of assignment, and keep in mind the typical (and non-negotiable) 3-year obligation (deployed joint assignments may reduce this commitment substantially) and potential education requirements to meet the qualification minimums.

So, let's call it like it is: your branch or career field manager probably won't be too excited or work too hard to set you up with a JDAL joint assignment if they don't feel you have the potential for continued growth or promotion. These slots are competitive and hard to come by, and Services typically won't send in a perceived catfish amongst the other Service studs, and possibly tarnish their own image, or burn that potential GO-shaping experience. Your record needs to be pretty strong for consideration, so if you keep getting the joint door slammed in your face, you may need to make some adjustments.

Equivalent credit. Many assignments within the joint arena won't appear on the JDAL, but this doesn't necessarily preclude you from attempting to solicit joint credit based on the particulars of your joint experience. Troopers serving in Joint Task Forces often fall within this category due to the 'temporary' nature of these organizations. Although the JDAL is updated on a regular basis, it rarely includes JTF jobs as a result. When considering whether to apply for joint credit, determine the impact your position held within that joint community (lead, evaluate, rate, span of influence, etc.), then complete your application on the Joint Qualification System (JQS) website, but don't expect a speedy points adjudication and response; it took my packet over 20 months.

This is my Boom-Stick. Once in the joint world, you may find yourself a stranger in a strange land. Be cognizant of your lexicon, as most of your Army terms, acronyms, expressions, and cultural norms will hold little meaning amongst your new, colorfully dressed, eclectic

Air Force Vegetables

peers. Also, be sensitive to Service jargon pitfalls; you might just instigate some butt-hurt looks with perceived disparaging comments (don't call Navy warrants 'Chief,' and never call a ship a boat—unless it's the week before the Army-Navy game). If you find yourself stationed on a base belonging to another Service, do some quick research on flag etiquette, and know how to render courtesies when the appropriate times come (e.g., there is a different interpretation of 'Retreat' on Navy bases than we're used to on Army posts, to include when the bugle call should be played and what actions should be taken).

Evaluations. When it came time for my annual evaluation while serving in the JTF, I solicited guidance from the Secretary of the Joint Staff (SJS) to determine when or if my bosses needed my support form and OER shell. I was stupefied when directed to simply write my own evaluation. "Rater and senior rater blocks?" I asked doubtfully. "Yes," came the hushed response. After further inquiry and reflection, it became readily obvious that my Navy bosses were not only strangers to the Army's evaluation system, they were reluctant to embark upon drafting comments that might not comply with accepted Army norms, and perhaps become detrimental. In consultation with some of my Navy and Coast Guard peers, it was also apparent that within those cultures, it was not unusual to draft your entire evaluation report, and merely solicit raters for concurrence and 'blocking.' I soon faced a similar dilemma when it was time for me to draft an evaluation letter for an Air Force Major for whom I rated. Besides the great work and extreme talent he possessed, I had no idea how to capture and format it in this letter in such a way the Air Force community would understand it. I had to lean hard on my trooper to help me 'mold' his evaluation into something that would make sense to his Air Force board brethren. By this point, I could completely empathize with my Navy bosses.

Army = planner. As a Soldier stationed among other Services in ad hoc joint formations, don't be surprised if they look to you to be an expert on planning and problem solving. Let's face it: out of all the Services, the Army has the market cornered on abundant problem solving doctrine and methods such as MDMP. Ever notice the similarities between the Joint Planning Process and the Military Decision Making Process? That's no mere coincidence. In many regards one problem solving process looks quite a bit like the other, and the Army spends an inordinate (tortuous?) amount of time training and utilizing them—perhaps more so than any other Service. Make sure you have it down pat.

Manage Your Own Career

NO ONE ELSE WILL or should care about your career as much as you. Don't expect the Army to manage it for you. Do the math: there's one guy/girl at Branch that manages your file and those of all of your peers and classmates; that's a lot of ORBs to scrub. Is it Branch's responsibility to identify that you have one less AAM on your ORB than in your DA Photo? It would be nice, and on occasion it happens, but that's an unrealistic expectation for most. It took me years to notice on my own ORB that someone had accidentally put 'PAR' as my birthplace instead of 'PA.' Although I'd like to someday, I have never set foot in Paraguay.

You are your own best advocate. Potentially, there's already enough working against you, so don't contribute further to your own demise. You have an inherent responsibility to know when your career milestones will occur, and to take appropriate action to ensure your information at Branch/HRC is accurate and up-to-date. Not doing so is tantamount to shooting yourself in the foot. Let's review the bidding:

- **Log on and Verify.** Ensure your latest OER is online and posted in your personnel file and your board file (ask your boss whether or not a Complete the Record OER is appropriate for you).

- **Three toes.** Nothing convinces a board of your sloth like an outdated, under-ranked, or poorly QA/QC'd photo. DA photos are your first impression and introductory 'handshake' to board members. Don't let a Captain in green Class A uniform photo go before the LTC promotion board. Strip off your regimental branch crests, fourrageres, unit-specific awards not earned in the unit and, for all you Cav and Infantry guys—sorry—but the crossed sabers and blue cords got to go. I also recommend a new Service uniform (ASU) photo if there have been significant changes to rank, badges, awards, waistline, or tour hash marks.

Get one completed before you deploy!

- **DA Photo Expression Faux Pas.** Be mindful of what you convey to board members with your facial expression. The following garish categories should be avoided:
  > 'Thumb in a light socket'
  > 'Please don't hurt me'
  > 'I'm a zombie'
  > 'I'm an arrogant prick'
  > 'I'm going to rip your head off and crap down your neck'
  > 'Count these chins of mine'
  > 'You're kind of hot'
  > 'Take note of my capped molars'
  > 'I'm sleepy'
  > 'You bore me, and I couldn't give a crap what you think'

I submit that the expression you would rather convey to board members might be: 'I enjoy what I do,' 'I like going to work,' 'I'm confident,' and 'I care.' Don't be afraid to smile in your photo (I would recommend against a full-toothed, gummy grin, however). Board members tend to look more favorably upon someone who looks like they are happy to go to work verses the look of a glaring, draconian hard-ass. So, save the stern, 'I will max you out' look for your command photo, and think to yourself while staring at the camera, whom would I rather work for?

- **Translate ORB hieroglyphics.** Spend some time with your S1 to make your Section IX—'Assignment Information' readable at first glance. With the exception of your 'Current' Organization and Duty Title, your S1 can edit all past assignment units and titles. Change HRC gibberish such as '4ARRGT05 HHT RECON' to a more digestible 'HHT, 5th SQDN, 4th CAV.' Spell out duty title acronyms like 'SGI' or 'OC' as well.

- **It's hard**—damn hard—to prepare for your board while in theater. Know when your boards will meet, and backwards plan to figure out what you need to get done before you deploy. This may include scanning in and sending awards, transcripts, and completed degrees that might not have hit your file as well, and updating your ORB to reflect that you are/were deployed.

- **Certify.** Once you are convinced that your board file is straight, don't forget to certify the contents. Board members will know whether or not you took the time to do so.

Create a "History of Me" book. If you haven't done so already, consolidate and copy everything from PCS/Airborne school orders to DD214s, AAMs and OERs, and make a 3-ring binder that contains it all. Burn it to a disk and back it up in a dependable (private?) cloud. Don't trust that every award (certificates and orders) and OER will make it into HRC, or that it will stay there. This product will help you immensely every time a board meets.

Some Parting Deep Thoughts...

BELOW ARE SOME of my random observations and self-evident axioms that have proven true and somewhat helpful to me over the years. Perhaps they can be of some benefit to you as well.

- Your self-derived farewell or change-of-command speech should never be about you.

- Sarcasm doesn't translate well over the radio.

- Don't even think about getting out of the Army until you have taken the guidon for at least a day.

- Don't forsake an opportunity to keep your mouth shut, especially in meetings or on the radio.

- Most people don't wake up in the morning saying to themselves: "I can't wait to screw something up today." In most cases, give Soldiers the benefit of the doubt.

- Humility is the most underrated of leader attributes. It serves as the best friend of objective introspection—that precious catalyst of innovation and evolution.

- If given or able to make the time, don't pass up an opportunity to learn something from anyone, regardless of rank or position.

- There are no stupid questions except most asked of a guest speaker in an auditorium full of your own peers.

- Simplicity is the most neglected Principle of War.

- Brevity counts in all things military except when talking with Soldiers.

- One man's envelopment is frequently another man's frontal attack.

- Don't get too enamored with technology; an E-tool, protractor, and five-ply paper may well become your closest friends someday.

- At the Battalion level, there is rarely enough time to wargame more than one course of action. Spend time fleshing out a single, solid COA, and, given the latitude to do so, branch plan the hell out of it.

- When briefing, it's always "we" not "I," unless taking blame for a mistake or shortcoming.

- Take every opportunity to tout your troopers in front of your boss, their other bosses, and their families.

- Never say no, nay, never to Force Protection dogs.

- Train the suck; practice fighting in degraded mode every time, even on staff.

- You're either in a learning unit or a dying one.

- As a staff leader, don't confuse your proximity to the command group with a transference of their authority or firepower.

- TRADOC assignments should be coveted and sought-after positions for successful leaders to selflessly give back to the Community of Arms.

- The Four Horsemen of Career Apocalypse: integrity mishap, mojo misapplication, government money misuse, and classified information mishandling, accidental or otherwise.

- Figure out how best to apply yourself for the benefit of your unit based on your rank, position, talent, and education; balance that against the absolute requirement to lead Soldiers through presence.

- Don't be too quick to render judgment on someone based on what they do or do not have on their chest, shoulder, or sleeve.

- Your worst enemy and that of your unit's, may well be your own damn ego.

- In meetings, people are most likely sitting there thinking about what's not getting done instead of focusing on what you're prattling on about.

- Don't be too hasty to vote the diverging thinker off the island. That weirdo probably has some pretty damn good ideas.

- It's tough to make a joke these days without offending someone; when in doubt, disparage yourself.

- When the inevitable mandate comes to 'do more with less,' be creative and inventive in training and operations with both personnel and resources, but don't cower when it's the right time to stand up and say, "We just can't do this, or do this well." (note that can't ≠ won't)

- Refrain from vocalizing your political views in the office, in uniform, in front of subordinates, or on open social media.

- Your reputation and professional measure springs from a comprehensive and accurate assessment created more so by your peers and subordinates than by your easily duped and often distant bosses.

- Yes, the plan rarely survives first contact with the enemy, but don't use that as an excuse not to plan.

- A command philosophy longer then a page rarely gets read, regardless of echelon (size 4 font doesn't count).

- Most of the world's problems can be solved on a collaborative dry-erase board.

- Strive and hope for the best in your career, but remain pragmatic. Branch or sequel plan for setbacks, and manage your own expectations carefully.

- In the absence of further orders, buy more toner.

Keep Things in Perspective

FROM THIS POINT on in your career your reputation will carry more weight than your pure performance. Yes, you must be competent, but you must be more than just good at your job. Being a team player, being genuine, seeing the big picture, shaping your environment, the ability to anticipate problems and manage risk will contribute dramatically towards your reputation as a Field Grade officer. That reputation will percolate up to the Brigade Commander, who knows from experience that reputation counts and will take it into consideration when deciding whether you will command a Battalion as a Lieutenant Colonel. The reputation you establish as a Field Grade, in or out of Key and Developmental jobs, will follow you for the next ten years of your career. Manage this as you fight the day–to–day fight to make your unit the best it can be.

Have no doubt that there are plenty of frustrations and hardships on the immediate horizon, but don't let that distract you from the bigger picture. Like everything in life, especially life in the Army, successful field–gradeship and soldiering as a whole require a careful balance of work, leading Soldiers, caring for family, and maintaining yourself. When asked, many officers confess that the most rewarding aspect of Company command was the time spent with and the privilege of leading Soldiers on a daily basis. As a Field Grade and staff leader, this opportunity still remains, despite the exponentially increased workload. Don't push it aside to make space for PowerPoint slides, online training, and IPRs. Integrate it into your daily battle rhythm, and you will find your time on staff to be just as rewarding as your cherished time in command. Have a blast out there, team, and good luck!

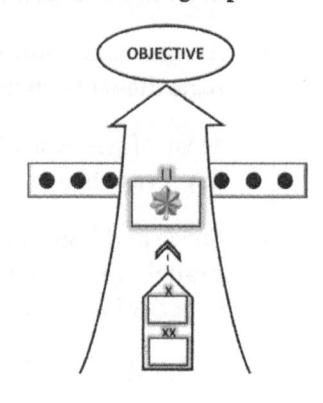

Epilogue

Culmination in the Breach—Hope, Disappointment, and the End of the Line

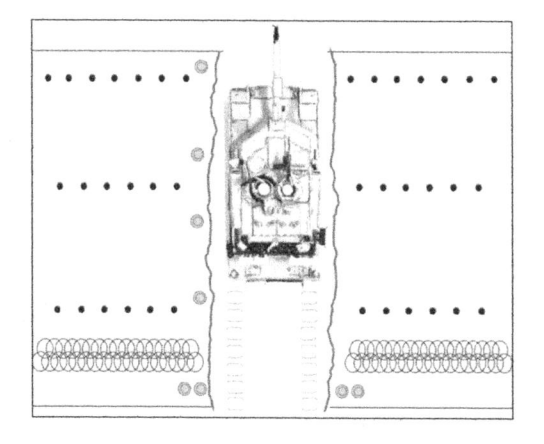

THE TRIGGER HAS BEEN MET. The conditions are set. The excitement of receiving that call to attack through the breach and seize the objective on the far side of the obstacle is nothing short of titillating. This is often dangerous turf, however. Haphazardly proofed, it now remains the biggest focal point of defending enemy direct and indirect fire systems, perhaps arrayed in a growing sense of nervous desperation, awaiting the tip of our spear as it blitzes through those last few meters of unmolested turf before the enemy is forced to displace or die in place. The lane beckons us on with tantalizing visions of grandeur, treacherously marked with flimsy cones placed by venerable sappers to show the nebulous line where the mines begin and end. The objective—that utopian goose egg arbitrarily drawn on the map so many hours ago, sits close at hand, now defended only by an abject few. It's within the grasp of our hastening tanks and Bradleys. But damn it, like an upstairs hallway in a horror flick, that breach can seem to stretch for an eternity. As you race through that narrow lane behind a cumbersome mine-plow, you can almost smell the ammunition cooking off inside those T-80 turrets when, much to your chagrin, it happens...

If you're feeling chipper because your career is on track and there are sunny skies as far as you can see, then there's probably little need to continue. If, however, that changes, or you already have doubts, or there are gloomy clouds clumping up on your horizon now, then by all means jam some Rage, grab a 40-dog, and read on...

It was just another Monday morning at the Command and General Staff College as I wove my way through the throngs of Majors choking the halls in an effort to reach my office before my first block of instruction. I passed by a couple of animated officers and couldn't help but pick up on their excited conversation that the Battalion Command list had just posted. It had been a few long months since my first command board had met, and, despite my angst, I had almost forgotten that the magic list would be released at some point. Apparently today was the day. Without pause I could feel my pulse start to race and a faint ringing began in my ears. As my pace quickened, I once again—perhaps for the hundredth time—mentally reviewed my file and weighed my chances of being on that list.

Slowly, however, the reality of the situation began to set in and, as my bottom lip started to poke out in an unknowing pout, my strides slowed with the sudden comprehension that I hadn't received any Sunday night congratulatory phone call. I was pretty sure that's how it went, but hey, maybe things were different as an instructor here at CGSC, or maybe I had failed to update the alert roster, just maybe... I tucked in my boo-boo lip and once again stepped it out, practically running up the stairs to my office. I quickly logged on, pulled up the HRC site and downloaded the command list. Starting from the top and with darting, buggy eyes, I hurriedly began to scan the names by command categories as the ringing in my ears and heart rate worked to a crescendo. I made it to the bottom of the list without finding my name, but thought that in my haste I might have missed it. Before going on to the alternate list, I began from the top once more, less franticly and scanning more deliberately this time, nonetheless, with a growing sense of unease. The outcome wouldn't change with a second scrub, however; I was nowhere to be found. As this realization started to take hold, my heartbeat subsided into dull and seemingly unfulfilling thumps, and the ringing in my ears faded to an irritating, distant whine. Remarkably enough, I did manage to find my name on the alternate list—a poor compensation prize, I thought, through a haze of bowel-twisting disappointment.

> But whom was I really kidding? I knew months—if not years ago that my file, although fairly strong, was still borderline at best. I had told myself time and time again that my chances were slim indeed, especially as I had refrained from competing in categories other than tactical commands, foolishly narrowing potential opportunities that much further. Yet, despite my best efforts to remain pragmatic and even pessimistic, I had secretly coddled and nurtured a growing hope that I would be selected for command, regardless of my suspect credentials. When I discovered otherwise, I will confess that it was tough to bear. I would have to get used to it however, as it would be but the first in a series of similar results to come. I'd like to say it got easier with every subsequent Dunphy-less list, but I'd be lying.

I debated whether to write about this sobering topic, but after some of my own subordinate Field Grades were faced with this kind of disappointment recently, I felt it might be beneficial to provide some frank and admittedly deeply personal perspective on the issue of major career setbacks, if for no other reasons than to recommend that leaders manage their own expectations accordingly, plan for it, and be prepared to assist their subordinates in doing the same. To be completely honest, I embarked upon the first edition of this manual through the course of my own dark introspection, having just suffered through the aforementioned, disheartening results of my initial command look, and trying to figure out where it (or I) had all gone wrong. As mentioned earlier, it would only be the first in a string of professional disappointments that would follow—subsequent command selection boards, Senior Service College selection boards, and finally my most recent O6 selection boards—all with similar, equally dismal, yet wholly predictable results.

Sixty percent of the time, you're selected every time. The Army faces yet again another restructuring effort in an attempt to make the most out of every dwindling dollar. In addition to the routine 'trimming of the fat' that comes with competitive selection, many more of us might find ourselves not making that next career milestone list. Some may even qualify for an abrupt and painful pink slip. Unfortunately, a graphical depiction of Army promotions over time looks more like an exponentially decaying function than a horizontal line. By the very nature and construct of our Service, there have to be cuts at every level, as obviously not every 2LT can attain the rank of General officer, nor every Private, a Command Sergeant Major. Historically, the more senior the rank, the deeper the per capita cut. Short of becoming the Chairman of the Joint Chiefs of Staff, at some point in your military career, a door will be shut in your face. It happens often enough through the results of a promotion, command selection, or school selection board (ILE, SSC, etc.), and it can be one of the most distressing episodes in a person's personal and professional life. Some of us face it earlier than we wanted to or were ready for, but most—if not all of us—will face it at some point, regardless of how good we are or think we are.

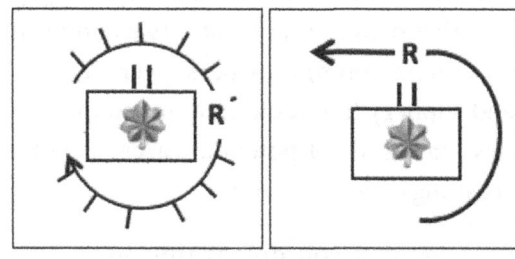

Board Outcomes?

Reading the tealeaves. Many of us will probably have some hunch or inclination of our impending demise; it's not like we don't have access to the same file that the board members do—our evaluations, records brief, DA photo, etc. If we remain objective enough, and with the help of mentors, bosses that give a crap, and Branch reps (that give a crap), we can probably develop and maintain a pragmatic and realistic picture of our chances for making the next grade or that elusive command selection list; call it 'Career IPB.'

The Good, the Bad, and the Ugly. There are probably three types of troopers facing any board: those that definitively know they won't make the cut (some due to derogatory or chronically poor performance), those that know they will without fail (damn overachievers), and the rest of us poor bastards. The last, and by far the largest category can be perhaps the most aggravating or unnerving to be in, as one must continue to toil in an environment of ambiguity, until that inevitable phone call or list comes out (for better or for worse) and then perhaps do it all again the following year.

Hope springs eternal...much to our own demise. So, for those of us trapped in the fog of selection uncertainty, hope becomes that which we cling to in the distant recesses of our mind, the thing that incites fantasies of the Velcro Tear, cutting rank-frosting cakes, and Ruffles and Flourishes, with the family sitting proudly at our shoulder. It is also that demon that most sets us up for the agony associated with the kick-in-the-gut/groin pass-over. Hope: the ever-present, unsolicited and slippery handhold that, no matter how precarious, simultaneously sows the seeds of potential career delights and of abject rejection and crushing disappointment.

So, what can you do? At this point in your careers, considering the prescriptions written in this document and your own continued efforts to make your units and yourselves better, frankly, quite a few things. I recommend that you maintain a realistic perspective on your chances of selection and advancement at every milestone, and coach your subordinates to do the same, especially if your scrutiny of their files might lead you to think they face similar circumstances. In so doing, you might avoid or at least mitigate the shock associated with such a perceived catastrophic setback.

Don't sell yourself short, Major, you're a tremendous slouch. But not really. Understand that, in many regards, your inability to get picked up for that next great school or rank may have little to do with who you are as a professional or as a person. Your identity and self-worth should or need not be based upon your career, or defined by its success or failure. One should not wholly render it a direct and dubious reflection of one's self or performance either. Rather, consider it as a necessary condition of the steep competition that marks our career choice, as well as the increasingly experienced and talented pool of leaders one competes against at higher levels. Also know that our selection and evaluation systems are not without their own fallibilities; oftentimes rater profiles, timing, and personalities may contribute to the fog of war and influence an otherwise relatively sound selection process. This should not be an excuse to dismiss or discard a lack of selection as nonsense, however. One should take notice, take pause, and consider this as an opportunity for further, deeper introspection to facilitate continued adjustment and betterment, or—if it suits you—departure and potential rebirth.

The plot thickens (or thins, rather)...

As fickle fate would have it, I was activated off of the Alternate Command list just days after the bleak, yet identical results of my second CSL board were posted. Although on relatively short notice, it was, in fact, a pretty great opportunity to serve in an Infantry Brigade's Light Cavalry Squadron—a dream job by any measure (despite an absence of tanks)—and one that, if executed well enough, might put me squarely back in the ring. However, the timing couldn't have been worse for my family, ironically as a result of actions taken earlier under the assumption of potential career culmination. I now faced a decision of enormous magnitude, with dramatic implications for both my family and my career. My mission analysis could only be described as torturous as, over the next 24 hours, I weighed the pros and cons of taking command or turning it down—a dilemma I never thought I would be lucky or unlucky enough to face. After a sleepless night consisting of much difficult deliberation, I grudgingly ended up choosing not to take command.

The following day I picked up the phone and made the toughest call of my career back to Armor Branch, informing them of my choice. It was the equivalent of sealing my own fate, as a formal declination (with prejudice) would preclude me from any other potential future command selections, and—certainly within Armor ranch—remove me entirely from any further consideration for promotion.

So, there it was: the Holy Grail of Career Redemption and I let it slip through my grasp to thwart potential undue family hardship. But, at the end of the day, you have to decide what's most important in your life, and pursue those priorities beyond mere lip service. I'd like to say that I never looked back at that moment and second-guessed my decision, but that too would be a lie. Every day I pull up my boot strings, I think to myself 'oh, but what if?'

Don't give up the ship. When the time comes, understand that an inability to achieve a goal need not represent career culmination, and it certainly doesn't constitute failure. There are subsequent 'looks,' other opportunities to excel in your base branch or otherwise, and years of potentially broadening, wholesome, and rewarding service with great Soldiers still ahead of you, regardless of the dimming chances for increased rank or position. How you define success and how you interpret or react to events such as these is largely up to you. You can become that disgruntled and embittered trooper that dourly trudges the halls, playing the role of the hapless and blameless victim; one merely trodden upon, churned up, and spit out by the impassive and soulless Green Machine. You can let yourself be overcome with apathy, willing to half-heartedly contribute to your organization in a 'Retired On Active Duty' (ROAD) fashion, content to draw a paycheck and burn oxygen. Or you can lament briefly, adjust fire, and then resume the attack once more—unto the breach. My recommendation is to protect yourself ahead of time by remaining objective, develop branch plans to cope with or overcome career set-backs, be prepared to work through the inevitable, and keep fighting; you might be surprised at what opportunities still await. This is, after all, a survival guide.

Fade to black. If you do find yourself leaning toward getting out or retiring, be cautious in sharing that inclination until you actually drop your paperwork. Certainly do the math beforehand in discussing this key decision with your trusted mentors outside of your current unit, but think twice about letting that cat out of the bag within your own headquarters until the conditions are properly set. I say this because you may inadvertently limit your options or shut doors that you might otherwise want to keep open in case the operational environment changes. The following anecdotes marked some critical inflection points in my career. The first was a good news story that resulted from abiding by this policy. The second, however, not so much. Maybe they can both serve as figurative career lighthouses for those of you facing similar dilemmas.

Sometime during my second year of Company command and while on deployment to Bosnia, I had pretty much silently resigned myself to getting out of the Army, mainly due to a growing sense of disillusionment at the time and an ebbing tolerance of tyrannical leaders. My Brigade Commander, a truly outstanding officer and one that I deeply respected, came to visit me one day at my Company FOB, and indicated that he wanted to influence Armor Branch into considering me for a position as a Small Group Instructor at the Armor Captain's Career Course. This was an honor to say the least, and one that I hardly felt I deserved or was qualified for. A large part of me couldn't care less, however, as I had already decided that my next assignment would probably be my last. Sure, I would go and do the SGI thing for a year, while working in concert to set the conditions before pinning on my 'Mr. Dunphy' rank. I wisely kept my pie-hole shut nonetheless and, upon my redeployment, I made my PCS move to Fort Knox and began work. Within several months, I had completely changed my views on my career in the Army. My experience was so positive and eye opening that, ironically, I decided before the end of that year that I was in it for the long haul, for better or for worse.

I often think back to my decision to hold my tongue, and I realize now that it was probably one of the soundest I would ever make (among a very paltry few). Who knows where I would have ended up had I not been an SGI? In all likelihood, I would have bailed out and missed all of the other great assignments, people, and opportunities I had over my subsequent and gratifying years of service. If you had cornered me at my outgoing change of command ceremony and asked me then where I thought I was going to be in ten years, the last thing I would have said was "still in uniform," despite my deep love of Company Command. I should have better committed that lesson to memory...

For tough reasons I've already vaguely alluded to, I declined command in the summer of 2011. As discussed, I was teaching at the Command and General Staff College (CGSC) when it happened. I knew what that declination meant, and those implications helped shape my penchant then to retire, and possibly join the Staff College's civilian faculty at some point, as I did enjoy teaching. I conveyed as much to the Brigadier General responsible for accepting my letter of declination, and in predictable fashion, grades soon reflected. My fall from grace was quick and dirty. I dropped from a relatively noble 'Ranked 5 of 110 O-5s I senior rate' (Above Center of Mass) the year before to a meager opening line of 'one of our best instructors' (read: yawn—Center of Mass). This despite the exceptional, enumerated numbers from other raters, and some accolades earned over the course of the year in, arguably, one of my better performances to date. But I could hardly hold it against him, as he probably didn't have to think too deeply about 'burning' another top-block on an officer that had no potential for future promotion—at least in Armor Branch—and one that had just told him he was thinking about retiring.

I deployed a short time later, and over the course of that deployment and through my interactions with my old bosses at CGSC, it became obvious that a confluence of gloomy budgetary events weren't going to facilitate a lot of opportunity hires upon my return and retirement. My plan was starting to fall apart. I realized that my prospects were grim indeed, and in a last ditch effort, I picked up the phone and shopped the functional area career fields until I found some poor suckers to take me in (I didn't bother calling those ORSA guys). I gratefully executed a rapid jump from Armor Branch into the open arms of Space Operations (FA40) in the hopes of redeeming my career, maintaining some relevancy, and perhaps remaining eligible for future promotions (a declination statement in the FA40 community shouldn't have excluded me from consideration). That last evaluation, however, would come back to haunt me when my promotion board met about a year later. Predictably I didn't make the cut, despite the presence of other consistent kickass OERs (I know because I had to write most of them).

I reflect on that interview with my senior rater, and realize that it too was one of those pivotal and life-shaping moments of my career. By conveying my tentative desire to retire, I put one more rather large nail in my own coffin. In hindsight, I should have thought that encounter through more carefully, and perhaps stated my intent to keep my options open, to lean forward in the saddle, and to serve with all due vigor, maybe even discussing with him the prospect of branch transfer and continued competitive development. Perhaps that would have stayed his hand somewhat, just enough to allow me to bow out with some grace and remain competitive in that other field if or when the time came.

Well crap. What can you do? So...there you have it, folks...my final, self-professed manifesto. There's a good chance that after reading this book you're wondering how I ever made it beyond the rank of Captain. In hindsight, instigated through the construction of much of this work, I often wonder the same thing. I genuinely hope this can be of some assistance to you in avoiding some of the many piles of career dog poo I managed to step in. Do I have regrets? Isaac Hayes said it best: "you damn right." Like all of us, I suppose. I'd like to think that I've learned from them and adjusted accordingly, but that doesn't stop them from clinging to me like tenacious, itchy ticks, burrowing ever deeper under the skin of my psyche. However, despite their nagging presence, I've had very few misgivings about my subsequent years of service, and I continue to strive to make a difference, grow, and enjoy what I do and with whom I do it with (almost) everyday, regardless of what rank resides at the center of my chest. As simple as they are, these criteria may be the only ones really necessary for me to define occupational happiness and professional success, just maybe...

On occasion, I catch the alluring scent of a large diesel engine on the streets of DC—nothing too sexy, just a passing dump truck or bucket loader—and I'm momentarily consumed by a wave of nostalgia. The sights, sounds, and smells of the motor pool, tactical assembly area, tanks in a wedge, and troops in formation—it all comes rushing back and consumes me for the briefest and joyous of moments. Like a fading dream upon waking, I reach for it, clutching frantically—to live it for just a second longer. It leaves me feeling a bit melancholy, but thoughtful about the awesome opportunities, decisions, and experiences that many of you get to face in your near future. I'll admit; it makes me pretty damn envious. Mostly these days, perhaps some of my last in uniform, it's the faint smell of ozone accompanying a copy machine as it churns through another slide deck that calls to me. It, too, brings back memories, perhaps a tad less glamorous. In retrospect it seems I've manned a copy machine as much as a weapon over the years, but I suppose there's goodness in both tools of warfare.

Targets up, team; engage and report!
Cheers,
–DWD

Glossary of Acronyms

1SG First Sergeant

AA Alcoholics Anonymous

AAM Army Achievement Medal

AAR After Action Review

ABCS Army Battle Command Systems

ACOG Advanced Combat Optical Gunsight

ADA Air Defense Artillery

ADO Army Direct Ordering

AG Adjutant General

AGMB Advanced Guard Main Body

AMC Army Material Command

AO Area of Operation

AOAP Army Oil Analysis Program

APFT Army Physical Fitness Test

AR Armor/Army Regulation

ARP Army Run-on Paragraph

ASAP Astronomical Spectator Apocalyptic Prophylactics

ASP Ammunition Supply Point

ASU Army Service Uniform

ATN Army Training Network

AWG Asymmetric Warfare Group

BBQ Seriously?

BCTP Battle Command Training Program (now MCTP)

BCT Brigade Combat Team

BDE Brigade

BFT Blue Force Tracker

BG Brigadier General

BLUF Bottom Line up Front

BMO Battalion Maintenance Officer

BMT Battalion Maintenance Tech

BN CDR Battalion Commander

BSA Brigade Support Area

BUA Battle Update & Assessment

BUB Battle Update Brief

C2/MC Command & Control/Mission Command

CA Civil Affairs

CAB Combined Arms Battalion

CALL Center for Army Lessons Learned

CASSS Combined Arms Services Staff School

CC Courtesy Copy

CCIR Commander's Critical Information Requirements

CCTT Close Combat Tactical Trainer

CDR Commander

CG Commanding General

CGSC Command and General Staff College

GLOSSARY 193

CHEMO Chemical Officer

CHOPS Chief of (Current) Operations

CIF Central Issue Facility

CIP Command Inspection Program

CO2 Consideration of Others

COA Course of Action

COB Close of Business

COC Chain of Command

COIN Counter–Insurgency

COMSEC Communications Security

CONUS Continental United States

COP Common Operating Picture/Combat Out–Post

CP Check Point/Command Post

CPOF Command Post of the Future

CPT Captain

CPX Command Post Exercise

CRP Combat Reconnaissance Patrol

CS 2–Chlorobenzalmalononitrile (uh... tear gas)

CSL Command Select List

CSM Command Sergeant Major

CTC Combat Training Center

CUA Commander's Update and Assessment

DA Department of the Army

DAGR Defense Advanced GPS Receiver

DCO Deputy Commanding Officer/Defense Collaboration Online

DCS Defense Collaboration Service

DECMAT Decision Matrix

DISCOM Division Support Command

DIV Division

DMZ Demilitarized Zone

DOD Department of Defense

DOG Damn Old Guy

DOIM Department of Information Management

DRASH Deployable Ready Assembly Shelter

DRMO Defense Reutilization and Marketing Office

DSCA Defense Support of Civil Authorities

DSM Decision Support Matrix

DTMS Digital Training Management System

DUIs Driving Under the Influence

ECOA Enemy Course of Action

EFMB Expert Field Medical Badge

EIB Expert Infantryman's Badge

EML Environmental & Morale Leave

EOF Escalation of Force

EPA Environmental Protection Agency

EXSUM Executive Summary

FA Field Artillery

FBCB2 Force XXI Battle Command, Brigade and Below

FG Field Grade

FLIPL Financial Liability Investigation of Property Loss

FM Frequency Modulation

FOB Forward Operating Base

FOD Friend of Dunphy

FOO Field Ordering Officer

FORSCOM Forces Command

FOUO For Official Use Only

FRAGO(RD) Fragmentary Order

FRG Family Readiness Group

FRSA Family Readiness Support Assistant

FSE Forward Security Element

FSO Full Spectrum Operations

FTA "Blank The Army"

FYI For Your Information

FYSA For Your Situational Awareness

GCSS Global Cheese Slicing Seminar

GFI " Good F(riendly) Idea" (Public Consumption Version)

GO/FO General Officer/Flag Officer

GPS Global Positioning System

HHC Headquarters & Headquarters Company

HHT Headquarters & Headquarters Troop

HQ Headquarters

HRC Human Resources Command

IA Information Assurance

IED Improvised Explosive Device

IG Inspector General

ILE Intermediate Level Education

INTSUM Intelligence Summary

IO Investigating Officer/Information Operations

IPA India Pale Ale

IPB Intelligence Preparation of the Battlefield

IPR In-Progress/In-Process Review

ISOPREP Isolated Personnel Report

ISR Intelligence, Surveillance, and Reconnaissance

IT Information Technology

ITO Installation Transportation Office

JAG Judge Advocate General

JDAL Joint Duty Assignment List

JPME Joint Professional Military Education

JQS Joint Qualification System

JSS Joint Security Station

JTF Joint Task Force

KD Key & Developmental

LD Line of Departure

LNO Liaison Officer

LOGPAC Logistical Package

LT Lieutenant

LTC Lieutenant Colonel

MCO Major Combat Operations

MCTP/BCTP Mission Command Training Program

GLOSSARY **195**

MDMP Military Decision Making Process

MEB Medical Evaluation Board

MEDPROS Medical Protection System

METL Mission Essential Task List

METT-TC Mission, Enemy, Terrain, Troops/Equipment, Time, Civilian

MG cMoanjsoird eGraetnioernasl

MI Military Intelligence

MIC Mechanized Infantry Company

MILES Multiple Integrated Laser Engagement System

MITT Military Training Team

MMRB MOS Medical Retention Board

MO Modus Operandi

MOPP Mission-Oriented Protective Posture

MOS Military Occupational Specialty

MRE Meal Ready-to-Eat (relative term)

MSR Main Supply Route

MTOE Modified Table of Organization and Equipment

MWR Morale, Welfare, & Recreation

NCOER Non-Commissioned Officer Evaluation Report

NCOIC Non-Commissioned Officer In-Charge

NIPR Non-secure Internet Protocol Router Network

ND Negligent Discharge

NSN National Stock Number

NVG Night Vision Goggles

OBJ Objective

OC(T) Observer/Controller (/Trainer)

OD Ordinance Corps

OE Operational Environment

OER Officer Evaluation Report

OIC Officer in Charge

OIP Organizational Inspection Program

OPD Officer Professional Development

OPFOR Opposing Force

OPORD Operations Order

OPSEC Operational Security

OPTEMPO Operations Tempo

ORB Officer Record Brief

ORSA Operations Research & Systems Analysis

PACE Primary, Alternate, Contingency, Emergency

PB Property Book

PBO Property Book Office

PBUSE Property Book Unit Supply, Enhanced (old)

PCS Permanent Change of Station

PL Platoon Leader

PLGR Precision Lightweight GPS Receiver

POL Petroleum, Oil & Lubricants

POSH Prevention of Sexual Harassment

POV Privately Owned Vehicle

PPT PowerPoint

PSD Personal Security Detachment

PT Physical Training

Q2 Second–Round Qualification

QA/QC Quality Assurance/Quality Control

QTB Quarterly Training Brief

QTG Quarterly Training Guidance

RAV Remission Assistance Visit

REAR–D Rear Detachment

RHIP "Rank Has its Privileges"

RIP Relief in Place

ROAD Retired on Active Duty

ROE Rules of Engagement

RFI Request for Information

RDSP Rapid Decision and Synchronization Process

RSVP Répondez S'il Vous Plaît (reply please)

RUFMA Rotation Unit Field Maintenance Area (NTC motorpool)

SA Situational Awareness

SATB Semi–Annual Training Brief

SATCOM Satellite Communications

SCPD Staff Choagie Professional Development

SDO Staff Duty Officer

SERB Selective Early Retirement Board (a.k.a., the axe)

SERE Survival, Evasion, Resistance, Escape

SGI Small–Group Instructor

SGM Sergeant Major

S&M You've been bad

SGS Secretary of the General Staff

SICUP Command Post Tent System (got me)

SIGACT Significant Action

SIGO Signal Officer

SIM Simulations

SIPR Secure Internet Protocol Router

SIR Serious Incident Report

SITREP Situation Report

SITEMP Situational Template

SJS Secretary of the Joint Staff

SOM Soldier of the Month

SOPs Standing Operating Procedures

SPO Support Operations

SR Senior Rater

SRP Solider Readiness Processing

SSC Senior Service College

STX Situational Training Exercise

SWO Surface Warfare Officer

TAA Tactical Assembly Area

TACON Tactical Control

TASC Training Aids and Support Center

TC Tank/Truck Commander

TDA Table of Distribution & Allowances

TDG Tactical Decision Game

TDY Temporary Duty Assignment

TEWT Tactical Exercise Without Troops

TF Task Force

TMDE Test, Measurement & Diagnostic Equipment

TMP Transportation Motor Pool

TOC Tactical Operations Center

TRADOC Training & Doctrine Command

TRP Target Reference Point

TTP Tactic, Technique or Procedure

UCMJ Uniform Code of Military Justice

USR Unit Status Report

UHF Ultra-High Frequency

VBIED Vehicle-borne Improvised Explosive Device

VIC Vicinity

VIP "Very Important Person"

VGT Overhead Projector Slide

VTC Video Tele-conference

VSAT Very Small Aperture Terminal (dish)

WTF Exactly!

About the Author

LTC DAVID W. DUNPHY enlisted in the Army as an Infantryman in September of 1987, and commissioned as an Armor officer from the United States Military Academy at West Point in 1993. He served at multiple Armor, Cavalry, and Joint assignments, and instructed at the Armor Captain's Career Course, Texas A&M University, and the Army's Command and General Staff College. LTC Dunphy continues to toil on staff as an operations division chief for the Joint Force Headquarters—National Capital Region at Fort McNair, Washington, D.C. If you're in the area, drop him a line—the cheapskate might even buy you a beer.*

Combat action shot of the author proofreading yet another slide deck.

Originally born in Philadelphia, LTC Dunphy spent his early life as a military brat and thus has little claim to a permanent home. He is the lucky father of three great kids: Kristina, Alek, and Dylan.

*Domestic happy hour specials only; no substitutes, and definitely no White Zin.